# KNOW THE RISK

# KNOW THE RISK

## Learning from Errors and Accidents:
## Safety and Risk in Today's Technology

*Romney Beecher Duffey*

*John Walton Saull*

**An imprint of Elsevier Science**

Amsterdam   Boston   London   New York   Oxford   Paris   San Diego
San Francisco   Singapore   Sydney   Tokyo

Butterworth–Heinemann is an imprint of Elsevier Science.

∞ Recognizing the importance of preserving what has been written, Elsevier Science prints its books on acid-free paper whenever possible.

**Library of Congress Cataloging-in-Publication Data**

Duffey, R. B. (Romney B.)
　　Know the risk : learning from errors and accidents : safety and risk in today's
　technology / Romney Beecher Duffey, John Walton Saull.
　　　　p. cm.
　　Includes bibliographical references and index.
　　ISBN 0-7506-7596-9 (alk. paper)
　　1. Industrial safety. I. Saull, John Walton. II. Title.

　　T55 .D815 2003
　　620.8′6—dc21                                                     2002026289

**British Library Cataloguing-in-Publication Data**
A catalogue record for this book is available from the British Library.

The publisher offers special discounts on bulk orders of this book.
For information, please contact:

Manager of Special Sales
Elsevier Science
200 Wheeler Road, 6th Floor
Burlington, MA 01803
Tel: 781-313-4700
Fax: 781-313-4882

For information on all Butterworth–Heinemann publications available, contact our World Wide Web home page at: http://www.bh.com

10 9 8 7 6 5 4 3 2 1

Printed in the United States of America

*Minimum: the smallest, least quantity, amount or point possible*

*Error: mistaken belief, fault, mistake, something done amiss*

*Risk: possibility or likelihood of danger, injury, loss, etc.*

*Theory: a general proposition, to explain a group of phenomena*

— Universal English Dictionary

# PREFACE

## ERRORS IN TECHNOLOGY: HOW SAFE ARE YOU?

We live in a technological world, exposed to risks and errors and the fear of death. The aim of our book is to show how we can learn from the many errors and tragic accidents which have plagued our developing technological world. These startling reminders of our mortality threaten us all, putting us at risk of death or injury. To understand the risks we face, how to minimize them and predict the future, we need to challenge you to think about "How safe are you?"

The book explores many headline accidents which highlight human weaknesses in harnessing and exploiting the technology we have developed—from the *Titanic* to Chernobyl, Bhopal to Concorde, the *Mary Rose* to the Paddington rail crash—and we examine errors over which we have little or no control.

By analyzing the vast sets of data society has collected, we can show how the famous accidents and our everyday risks are related. We look at:

- Traveling in safety
- Working in safety
- Living in safety

The reader is led through the extensive data we have collected, systematically examining each of these areas (see contents) using data and examples of actual events and illustrative narrative. We discover how the common threads of the sequences and the differing confluence of human actions link these errors with machines.

Throughout human history our errors have plagued us, and as our technological world develops, errors persist. Fortunately, we learn from our mistakes, correcting and off-setting them by reducing the rate at which they occur. However, it appears that there is a limit below which the rate of errors cannot be reduced further, although their severity may be reduced. Our analysis illustrates how we are faced with a seemingly impenetrable barrier to error reduction.

A unifying concept and theory shows how errors can and should be analyzed so that learning and experience are accounted for, and we show, by using a Universal Learning Curve, how we can track and manage errors to reduce them to the smallest rate possible.

The fundamental observation is made that the rate of error reduction (learning) is proportional to the number of errors being made. *The learning rate depends directly on the number of errors that are occurring and is dependent on the accumulated experience.*

We stress the importance of a *learning environment* for the improvement of safety and of the correct measures for the accumulating experience. We prove the strength of our observations by comparing our findings to the recorded history of tragedies, disasters, accidents, and incidents in the chemical, airline, shipping, rail, automobile, nuclear, medical, industrial, and manufacturing technologies.

Several years have been devoted to this quest, gathering data and analyzing theories relating to error reduction, design improvement, management of errors, and assignment of cause. The analyzed data relate to millions of errors, covering intervals of up to some 200 years and recent spans of less than 10 years.

The approach, which we have named DSM (the Duffey–Saull Method), has a broad and general application to many industries, as well as societal and personal risks. Using the collected and actual human experience, it refutes the concept and search for "zero defects," placing both quality and safety management in the same learning context.

We want to save lives and inform. During our search for proof of the common thread we talked to many experts, looked at many industries and technologies, and kept questioning ourselves about what the data were telling us. Our aim is to challenge our readers to take a different look at the stream of threats, risks, dangers, statistics, and errors that surround us, by presenting a new and different perspective.

We can all contribute to the reduction of errors and the improvement of safety for the benefit of mankind.

*Romney B. Duffey*
*John W. Saull*

# CONTENTS

# DEDICATION AND ACKNOWLEDGMENTS

This text is dedicated to learning.

We are indebted to our wives, Victoria Duffey and Silvina Saull, who made all this work possible by provoking the original research on travel and safety, and who helped and encouraged our work over these many years. We are grateful to all those who have gathered, provided, and supplied their data to us, those who have shared their accumulated experience with us, and those who believed in the work and also those who did not.

We acknowledge those many friends and colleagues who have read chapters and fragments, and who have criticized, challenged, and encouraged us. We thank particularly the following partial list: Ron Lloyd on medical errors; Dan Meneley on reactors; Victor Snell on safety and criticality; Terry Rogers, E.A. McCulloch, and Patt McPherson. We also acknowledge specifically the inspiration and reinforcement supplied to us by the pioneering and visionary work of Karl Ott on learning curves, which we learned about very late in the day; and by Dale Bowslaugh's work on significant event data bases; David Yeoman's views on failure theory; Naoëlle Matahri's studies on latent errors; Bill Browder and Peter French's key railway studies; Richard Morgan's compilations of pressure vessel defects and failures; Danielle Beaton's discovery and analysis of water treatment and other event data; and Colin Allen's data and dialog on safety trips. We are also indebted to the organizations listed in the References, Bibliography, and Sources for their kind assistance in providing historical incident/accident data.

We are indebted to the Internet and special computer programs, without which this work would not have been even remotely possible. Its amazing gift of open data access and communication we have unashamedly and gratefully used to the full, for data acquisition, source research, and numerous downloads. We thank all those individuals and organizations that have made their information, reports, and data accessible and available over the Net. Many of the illustrations are from the public open reports of government agencies, whose excellent accident information is freely and openly available in electronic form.

We also thank Judy Charbonneau for her organizational skills and contributions; and Kelly Donovan, whose support was critical to this book's publication.

# ABOUT THE AUTHORS

The authors have collaborated on this text for more than 5 years, and it is the fruit of much original research and extensive worldwide experience. This unique work is the direct result of the long and active synergism between the two authors in two distinct but related fields of safety. Both authors are skilled lecturers and presenters and have executive and corporate-level experience and communications skills.

*Romney Duffey*, Ph.D., B.Sc. (Exon), is an internationally recognized scientist, manager, speaker, poet, and author, having published more than 140 papers and articles. Educated in England, he has more than 30 years of international experience in the United Kingdom, the United States, and Canada in the fields of technological safety and risk assessment, system design, and safety analysis. He is a leading expert in commercial nuclear reactor studies. After a distinguished career working in the U.S. utility industry and for U.S. government agencies, and leading a not-for-profit foundation, he is presently the Principal Scientist with AECL (Canada). He is a member of the American Society of Mechanical Engineers and past-Chair of the Nuclear Engineering Division, an active member of the American and Canadian Nuclear Societies, and a past-Chair of the American Nuclear Society Thermal Hydraulics Division. He has organized and chaired numerous international conferences. He is active in global environmental and energy studies and in advanced system design, and is currently leading work on advanced energy concepts. He is married with two children.

*John Saull*, C.Eng., FRAeS, DAE, is an internationally known aviation regulator and author, having written many technical papers on aviation safety. He has more than 40 years of experience in commercial aircraft certification, manufacturing, operations, and maintenance, and he is a leading expert in safety management and human error. He has held a pilot license for 30 years and co-owns a light aircraft. He is presently the executive director of the International Federation of Airworthiness (IFA), after a distinguished and fascinating career with the Air Registration Board and the U.K. Civil Aviation Authority, where he was chief surveyor and head of operating standards.

He spent 7 years on the development of the Concorde supersonic transport aircraft, responsible for the construction standards. He also held positions overseas, in Argentina and the West Indies. He is a council member of the Royal Aeronautical Society, a Freeman of the City of London and Freeman of the Guild of Air Pilots and Air Navigators, and Chairman of the Cranfield University College of Aeronautics Alumni Association. He has organized and chaired numerous international industry-wide conferences and meetings. He is married with three children.

# KNOW THE RISK

# INTRODUCTION: ARE YOU SAFE?

*"Accidents happen. In fact, accidents are what life is about."*

—Tom Peters, "The Pursuit of Wow"

When asked how safe are *we*, we usually think of whether we might be run over, or shot, or robbed. We do not usually think of the technological world that surrounds us as a safety hazard or as able to hurt us without our knowledge. If it does hurt us in some way, we are surprised and ask the "authorities" to make sure that whatever it was it couldn't happen (to us or anyone else) ever again. We ask that whoever is responsible for doing it "to us" be punished, chastised, or prosecuted. After all, it was not our fault.

Or was it?

We would all like to be safe, to be free from the fear of death or injury, and able to live in the modern world safely and without fear. When we ask, "How safe are you?" there are really two meanings to this question.

The first meaning or context is really your feeling for your own *personal safety*. Are you safe when you go about your life in today's technological world, using a car, working a job, crossing the street, traveling to a relative, flying on vacation, or visiting the doctor? Are you subject to the risk of death or injury from the errors of others, and can mistakes made by others—failing to brake, misdiagnosing your illness—kill or injure you? And who and what are to blame?

After all, lawyers make a profession and a living from assigning blame for accidents and injuries, and so do we all as humans. Others make a profession of safety and managing and applying the "culture" of safety systems.

1

The second meaning of the question, which is equally profound, is really your own *social safety*. Are you safe in what you do yourself when you make mistakes in your job and work, in operating a machine, in coding your software, or in writing rules and practices for others to follow? Or are you perfect? Do your errors contribute to the overall safety of society? Do you put others at risk in what you do or by the mistakes you make? Are you at risk yourself from dismissal or reprimand for making an error?

Nothing and no one is perfect. Of course, we all make mistakes—but when the errors can kill, injure, or maim, or destroy a city, as humans we agree that this is serious stuff. We all make errors at some time, but do some humans make more than others do, and do some errors occur more often? Are they avoidable? Why are some more important? Can we predict error rates and prevent them? Can our vast set and array of modern safety systems, laws, regulations, inspectors, and courts prevent errors? Can errors be eliminated or reduced to such a level that they don't matter? Are the errors an integral part of our use of and access to modern technology?

As we rush headlong into a world run by computers and technology, surrounded by the products and processes of the industrialized world, the key question is: When are errors "acceptable" or at least tolerable? Newspapers and safety authorities, consultants and lawyers all make a living from reporting, regulating, investigating, and prosecuting lapses of safety discipline and errors of thought or deed. The more the error is intended, blatant, or of great consequence, the more likely the sensationalist reporting and the more public and political the clamor for inquiry, retribution, and revenge.

We ourselves have struggled with answers to the question of errors in modern high technologies. This led us to search many industries for data, to read many textbooks on the culture, psychology, and cause of accidents, and to conduct many new analyses.

*We wanted to find out if it was possible to measure, predict, or learn about the errors, accidents, and risks we are exposed to.* We, as authors, also have a real personal interest in this topic, which intersects and complements our social interest. One of us deals with the safety and performance of airplanes in which we also travel; the other deals with the safety and performance of nuclear power plants on which we rely for energy. We have talked to experts and experienced the culture that exists in today's high-tech world. We found that modern society is committed, even dedicated, to the collection of error data, such as death and injury statistics. These data are often expensive and difficult to obtain and are used by insurers, by judges, by regulators, and by ourselves to form ideas about risk and the probability of injury or death. But no one is really tracking what *all this data means* and how it all might be interrelated, since everyone is to some degree inevitably locked inside his or her own specialty, field, or paradigm.

As a modern industrialized society, we collect and publish accident data, we track and measure the consequences, and governments and agencies track and publish the information. Industry tracks their own data on errors of all kinds, in operation and manufacturing, as well as on deaths, accidents, and injuries. There is a plethora of numbers, and the World Wide Web has made much of this instantly available to us all. We are grateful to all those whose work and dedication have made it possible to compile, analyze, and interpret this veritable "data mine."

## HUMAN ERROR IS PART OF THE RISK

We all are committed to error reduction and improvement in our personal and professional lives. After all, a safe world must be a better place to be in. We want to travel, work, and play in safety, and even in high-risk recreation we usually try to take as many safety precautions as possible. We accept the risk of some daily activities (driving to work, catching a train), and these are *voluntary risks* we take. Other risks we do not expect to happen—our doctor or hospital making a mistake, or a chemical plant or fuel tanker exploding in our neighborhood, or a plane crashing nearby, or a flood affecting us. These are *involuntary risks.* Much has been written on distinguishing these two. What actually happens is that we accept the everyday risks where we have to, and we ignore or exclude the others as too uncomfortable or deadly to contemplate as we go about our lives.

We delegate the responsibility of reducing all those accidents and errors to others— regulators, safety experts, licensing bodies, standards setters, lawyers and courts, and politicians. They strive conscientiously to reduce the cause of accidents and errors to the minimum that is socially acceptable.

The public perception of safety is largely driven by media reports and opinions. It usually targets public services, for example, railways and the medical profession. In 2000 the U.K. railway system suffered a large downturn in passengers following two fatal crashes in the space of a few months. Following the second crash, the system was required to apply hundreds of speed restrictions due to the poor condition of the rails, which had to be replaced. This led to major inconveniences, as well as concerns for safety. The safety record was in fact not bad by international standards. It was perceived in some quarters that this degraded rail situation was the direct result of privatization of the railways, which separated the rail track network from the train operating companies.

The medical profession in some countries has come under increasing public scrutiny following reports of misadministrations during operations and questionable practices involving errors during the care of patients. The large numbers of apparently unnecessary deaths have been challenged. The healthcare systems in these countries are under severe public pressure to deliver acceptable patient diagnosis and care standards.

We are interested in determining whether we are learning from our mistakes and whether there is a minimum achievable rate for human error. Hence, there could be a rate below which error cannot be eliminated in a system or design in which there is human intervention or reliance. That rate is despite all the best efforts to reduce it further. This simple question has significant importance in complex technological systems where a human, as an operator, designer, or producer, can make sometimes fatal—or at least very expensive—errors that lead to loss of life or production. These mistakes are usually characterized as errors of omission and commission, and many studies have examined the contribution and quantification of human error rates. The error may occur during manufacture, operation, or maintenance, or in the use or the process itself. It may even be embedded in a flawed design.

The professionals in this field of human error study human reliability analysis (HRA), which includes what is called the man–machine interface (MMI) or human–machine interface (HMI). They examine the effectiveness and layout of the design of the work-place and its functioning. They study the design of dials, switches, and knobs; the messaging and alarms and the use of video and computer screens; the work flow and work load during operation and emergencies; the actions and reactions of the operating team; the value and type of procedures; and the theoretical and mental models for the causes of error. This type of analysis then allows an assignment or quantification of the human error probability (HEP). The HEP is a function of the actual tasks undertaken, the stress and time available, the availability and quality of training and procedures, the correctness and effectiveness of team decisions and actions, and the degree and extent of the management support structure, including the rewards and punishments for the actions. *Thus, all actions have to be viewed in what is called the context, meaning the environment or circumstances, in which they occur.* Human-centered design (HCD) considerations are taking increasing prominence in an attempt to reduce the possibilities of human error during in-service operation.

In practice, the chance of error is found to be a function of the time available for it to happen, the degree to which a problem or error may be hidden (a so-called *latent error*), the extent of the contextual influences (so-called *performance-shaping functions*), and the degree to which corrective actions occur (so-called *recovery actions*). *Errors are often found to be of multiple causation, with many factors acting in differing combinations contributing to the outcome and the sequence of actions and events.*

This rich literature is accompanied by the extensive inquiries and investigation of accidents, which, by definition, concentrate on the observed or observable errors. Accident investigations focus specifically on determining cause. For highly reliable systems we are at the limits of observability, since errors are often rare and may have complex sociological as well as psychological causes.

The determination and apportionment of blame is the specific function of the legal system. Legal action is progressing, aimed at "corporate manslaughter" convictions in high-profile accidents. We have noted that high-ranking companies have frequently been given immunity from prosecution in return for the inside story. Whistleblowers, although often protected by the law, usually face internal pressures and harassment because of their disclosures. In a true *learning environment* none of this would happen.

It is often difficult to prove who is directly responsible for serious errors and organization-fatal flaws. They might, and often do, involve a whole management structure. In some industries, there are national requirements to have an accountable manager, often the chief executive, who has responsibility for the overall safety of the approved organization, usually those involving technology that directly affects the public—for example, commercial aviation and nuclear energy. This is part of the growing awareness that safety management is all-important to the protection of the operation.

A true *learning environment* is both reactive and proactive, not purely reactive, waiting for disaster to occur before taking action as is so commonly the case. This is really what is called an "embedded safety culture."

The treasure trove of psychological literature looks at the error process in the human mind, via the errors of commission or omission and the rule, skill, or knowledge base applied to the task at hand. Mechanical error rates have been observed and tabulated for turning a valve, reading a sign, interpreting a rule, and so on. There are frameworks and models for error causation, and concepts related to the multiple barriers are put in place to reduce errors. But this vast body of work either does not have a number in it for the error rate or at the other extreme has a series of unrelated error rates from individual tasks that must be pieced together like a human jigsaw puzzle.

We questioned whether the vast amount of existing and available data that we humans have now collected and recorded:

    (a) Show any systematic evidence of learning and of a minimum error rate
    (b) Provide insight into the numerical value for such rare occurrences
    (c) Indicate the potential contribution of human error in modern technological systems
    (d) Correspond to any theoretical or predictive treatment
    (e) Can be related to the observed factors discussed above

We chose examples of industries and activities, not just because we were familiar with them and because data are available, but where the system is complex with significant human intervention and there is a premium on safety and error-free performance. Accidents make good newspaper headlines and may be sensational. But are these avoidable? How many are simply the consequences of modern society indulging in high-risk activities, with the expectation that they are safe when in reality they have real risk?

Many such industries (e.g., the airline, nuclear, shipping, chemical, and rail sectors) have been the subject of considerable scrutiny and regulation on a worldwide basis. The inference from the *average or instantaneous* fatal accident rate is often that it appears to exhibit no further decline over the recent past, despite the best efforts to reduce it.

This has raised the possibility of an unacceptable future situation, in terms of relative risk, as the number of accidents rises with increasing activity, thus implying that further rate reductions are both needed and obtainable. We expect that, whether the hypothesis is true *or* false, the finding has potential significance in many fields (manufacturing, systems, design, etc.) and indicates how safely and reliably we may be able to design and operate complex systems. We should also be aware that outdated technology does not usually have the benefit and protection of updated design manufacturing and operating techniques. However, experience is an important factor in their overall safety of operation. There are many occasions where mandatory modifications and inspections are implemented to correct identified serious problems.

We also sought other similar trends embedded in the available data for many technologies, to see if the hypothesis was confirmed. After all, it is evident that decreasing the number of deaths due to errors is also highly desirable, despite the apparent and stubborn near-constancy of the many accident rates in recent years.

In practice, there is undoubtedly a limit to the financial resources that can be put into ensuring improved safety; it is a question of assessing where it is best to apply the available resources. Cost–benefit analysis is a part of everyday safety control but is not often understood by the general public.

## A World of Defects and Errors

The resolution to this question has a bearing on the goal and search for "zero defects" or minimum error rates in skill-based processes, and hence also on the commercial advantage in product liability and reliability claims. It is a serious question as to whether, for example,

- Manufacturing defects can be claimed to be random when it is possibly an error inherent in the manufacturing process itself
- Certain transport modes represent the minimum financially acceptable risk for the observed error rate
- Medical administrations and practitioners can be held liable for errors in procedure or diagnosis

*We define an error as the unintended action or the combination or confluence of actions, equipment, and circumstances that causes an accident or injury in a technological system.*

Thus, whatever combination of actions, failures, and mistakes that occurred constitute an error. The error(s) can be in design, operation, or process, can be active or purely passive, can be single or multiple, can be spread over time, and can be committed by more than one person. The error can penetrate, work around, defeat, or avoid multiple safety barriers and systems, and this can be by default or deliberate action.

## Accidents That Cannot Happen

Humans are very clever. We tend to believe we are infallible, only to be reminded time and again that we are "only human." The greatest achievements of the modern age have taken immense efforts, sacrifice, and risk: landing a man on the Moon; traveling faster than sound; splitting the atom; discovering North America; climbing Mount Everest; developing cancer treatments and nuclear resonance imagery; designing the machinery of the Industrial Age. All these discoveries and efforts are not without errors that have happened on the way, measured as those who have died in the attempt or the process, or who have contributed their lives, as well as their brains and knowledge, to make it all happen or be possible. Just two examples are the Nobel laureates Pierre Curie, the pioneer of radiation, and Enrico Fermi, the leading nuclear physicist, who both tragically died early as a result of inadequate early knowledge of the effects of radiation from their experiments on themselves. Pioneers who have given their lives in the quest for discovery exist in every field.

*The path to the present is littered with the mistakes and errors of the past, most of which we did not foresee or simply did not believe could happen.*

We do not concern ourselves with the complex and empirical science of the psychology of errors, or the organizational, management, and social issues surrounding errors in society, cultures, and systems. We prefer to use the expression *learning environment* to describe the desirable characteristics for error reduction. It is sufficient that the resulting consequences were not intended, arise from interactions of the technological system and the humans in it, and do not discriminate between normal or abnormal errors. We define an error as something that has a consequence that results in some measurable effect, thus:

Error(s) ➜ Consequences ➜ Events, accidents, losses or injuries, deaths

Some of the consequences over history have been quite spectacular and very costly. The well-known list of headline-making and newsworthy events emerges, covering many centuries. This list includes many famous names and infamous events:

- ➜ the mighty wooden warships the *Mary Rose* and the *Vasa*, leaving port and sinking almost before their monarchs' eyes
- ➜ the burning of the great airship the *Hindenburg*, as it docked after a long flight and the movie cameras rolled
- ➜ the sinking of the great ocean liner *Titanic*, with many passengers' ice cold deaths
- ➜ the nuclear plant accidents and the ensuing panic at Three Mile Island, and from the fire and burning reactor at Chernobyl
- ➜ the major technological aerospace disasters of the explosion of the *Challenger* space shuttle and the flaming crash of the Concorde, both in full and public view
- ➜ the deaths from poisons explosively released from the chemical plant in Bhopal and the radiation at the JCO plant incident in Japan, both deadly but unseen
- ➜ the tragic Paddington train collision and the loss of the Air Alaska airliner, taking innocent passengers to their deaths
- ➜ the public water well contamination by deadly *E. coli*, causing deaths at Walkerton from just drinking the water
- ➜ the terrorist attacks on the World Trade Center and the Pentagon

There are many more: every day new incidents burgeon somewhere. These are the unthinkable accidents, which still occur worldwide, to the best-designed, best-run, most expensive and visible of human technological achievements. *It seems almost as if we are running a series of accidental experiments, but this time on ourselves.*

## WHAT DO THESE CATASTROPHES TELL US?

*They all share a set of common features: all were unexpected, all were not supposed to happen, and all transgressed safety rules and measures. They were and are all a confluence of unexpected events, errors, and circumstances, or a unique combination of unforeseen causes that we only understood afterwards.* ***They were preventable.***

A common thread also links this set of apparently disparate events: they all arise from the contribution and consequence of human error and its interaction with a technological device. They have all led to desires and rules designed to make changes, so they "cannot happen again." But of course, such accidents and errors still happen, which is all part of our in-built Darwinian human learning process. Not exactly the same events, or causes, or sequence, or result, but always a new set or *confluence of circumstances*, sometimes leading to a common cause within a specific technology.

*In a sense, the errors are inevitable: it is only the likelihood that we can change by learning from our mistakes.*

## A WORLD OF EXPECTED ACCIDENTS

After this set of bestselling headliners, compelling though it is, the list is more mundane and everyday—but still impressive, with major explosions, crashes, fires, and floods, some of which reach the national news headlines. Every day, too, there are many other accidents that are even more mundane and commonplace.

These accidents *are* part of the risk of being alive. Deaths and injuries occur from surgery, or drug misadministrations, from industrial accidents, from auto accidents, from ships sinking, and at railroad crossings. The world is full of death, injuries, and accidents, wrought by humans on our fellow humans and accepted as part of the risk—inevitable and known—of being alive, or trying to stay alive, in today's world. They all arise from what we state is an error, and share the common element of human error, somewhere in their cause.

## ALMOST COMPLETELY SAFE?

Some areas of endeavor pay special attention to or have a premium on safety, especially those of high risk or large potential consequence. Thus, while meeting the expected and regulated norm on industrial safety or design codes and practices, additional safety mechanisms, training, procedures, maintenance, and defect/incident reporting are implemented. These tools and approaches are in the "almost completely safe" activities, which—often because of the potential hazards or risk—must be very safety conscious. Examples are from the airline, nuclear, and chemical industries, so we examine their data with special interest—particularly because they are also our areas of expertise and background. We were motivated to find out what is possible in one industry or activity, and how applicable the experiences and practices are to others.

We examine and discuss the "headline stories," the catastrophes and the tragedies. We relate them to the more mundane everyday accidents and errors. We sought available data worldwide on injury and accident statistics in many fields of technology, and took a more holistic and combined empirical (looking for trends) and theoretical (failure rate analysis) approach. We rely largely on public data and analyze the existing accident rates. We seek evidence of trends in the accident rate and the number of accidents. Initially, we do not subdivide the data or attempt to distinguish among countries, manufacturers, the age of equipment, designs and regulations, and standards of operation. Thus, we are determining the overall trends for an entire service industry, independent of detailed national or commercial differences. *The common elements are simply technologies where the human is involved in the design, operation, or performance of the technological system.* For road transportation, we took a similar approach, seeking out the fatal accident data for a large population of automobiles and drivers. We were interested to see if these exhibited the same trends, and again we did not subdivide the data by manufacturer, make, or origin.

In addition, we examined data for industrial errors and accidents, to see if there was any evidence of similar trends in completely different technological and management systems. This addresses the question of the effect of safety culture on error rates, and the role of management, punishment, and the courts in error reduction.

As to the theory of errors, there is a gap between the psychological analysts, who are not necessarily numerically literate, and the reliability theorists, who are but who do not understand the human psyche. Both inhabit this arena of errors as their own specialty. We had to make some progress here, so we used ideas from a field known as "probabilistic safety analysis," an area of mathematical analysis introduced to quantify and predict actual accident rates using equipment and human reliability data.

We found it possible to bridge this gap between the mental models and mechanical reliability with a simple hypothesis: ***namely, that humans learn faster when they make more errors, and that reliability theory can be used to describe the overall error rate.***

## LEARNING FROM OUR MISTAKES

Thus, we learn from our mistakes, and it should be obvious that we do so somehow. This may be one manifestation of Darwinian natural selection, since those who do not learn are less likely to survive intact.

*So humans follow or obey a classic learning curve,* which inexorably trends to lower error rates; and thus accident and error rate statistics follow a similar trend when the *accumulated experience* is used as the basis for the measure of the activity. All humans and all technologies pass from novice to experienced status, from being new to becoming proven, from being a student to becoming an expert, where the rate of errors falls with increasing experience and knowledge. But the final, equilibrium, or ongoing error rate is not zero, even when a large experience base has been accumulated.

We do not like to admit that we make mistakes and commit errors: no one does. This new paradigm of learning causes problems with traditional error, death, and injury reporting and challenges the basic premise that errors occur and can be measured on a calendar and calculated on an instantaneous basis. The problem is that a technological system does not care about the calendar: its technological, often silicon-based clock ticks all the time, with an arbitrary beginning and an equally arbitrary end from when it is turned on to when it is turned off. Thus, when we present data and error rates plotted in this new way, it causes confusion to some. We will try to help avoid that problem.

*We point out that large bodies of historical error data are needed, which are systematically collected, defined, and analyzed on a common basis, to provide clear trends and estimates.*

Note that the accumulated experience can take many forms, but that the right parameter must be chosen, based on common sense. If the accumulated experience is for the operating equipment, then it is an accumulated operation. If it is for ships sinking, it is the number of shipping years; for planes the number of flights or flying hours; for employee accidents, the number of employee hours; for automobiles, the traffic density. Choosing the right parameter for the accumulated experience is as important as the error measurement itself.

## MODELING AND PREDICTING EVENT RATES

It would be desirable to determine and to actually predict event rates, and therefore one needs a theory.

It is common to fit trends in hazard (error, accident, or event rates) to arbitrary distributions or "models" for the hazard (error) rate. Rather than make arbitrary fits (polynomials are common) to these data, we have shown that according to our new approach, the DSM, the decreasing trend is due to the existence of a classic learning curve. This occurs because of the predominant contribution due to human error.

The fundamental assumption is that the rate of error reduction (learning) is proportional to the number of errors being made. *The rate of decrease in error (event) occurrence is proportional to the event rate.*

We will show the wide use of this simple but obvious assumption many times in fitting and predicting event data. Beyond the inherent simplicity, the advantages of using this learning curve (or resulting exponential form) are:

(a) It is theoretically based and therefore plausible
(b) It enables *predictions* and the associated confidence limits to be assigned based on historical (time history) event data
(c) It is not an entirely arbitrary choice, being dictated by the extension of reliability theory to human errors and accidents

To analyze the data according to the DSM, *the accumulated experience* can and should be used as the explanatory variable. Thus the new DSM approach provides a very different view of the data using the accumulated experience and providing a ready comparison to the constant rate (CR). Thus it is clear whether a learning curve is being followed or not, and the instantaneous rate (IR) and the accumulated rate (AR) can show quite *opposite* trends for very good reasons.

Thus the DSM provides:

(a) Comparison to the inherent expectation of constant rates
(b) Prediction of future event rates based on the accumulated experience
(c) Examination of the IR (instantaneous error [event] rate), which gives the trend
(d) Comparison of AR to CR
(e) Determination of whether the data set is well behaved and/or follows a learning curve
(f) Quantification of the minimum error rate that has been attained
(g) Theoretical basis and underpinning for the above prognostications

The theory does not pretend to disentangle or explain all the fundamental behavioral and psychological contributions. It is an "emergent theory," whose underlying complexities combine and interact to appear as an orderly trend that is fundamental in and by itself.

We may naturally include the effect and consequences of another human attribute, namely forgetting. It is evident that the rate of human learning is not perfect and that errors can be caused not only by doing something wrong, but also by forgetting to do something. This is the fundamental distinction between "errors of commission" (taking the wrong action) and the opposite "errors of omission" (not taking the right action), which are the descriptors used in the science of human factors.

If we do not mention it for a while, it may simply be because we have forgotten!

## THE HUMAN CONTRIBUTION

What is the role and contribution of the human in causing errors and accidents?

This is usually not measurable and is embedded in other factors related to the cause, such as inadequate training, inattention, poor procedures, errors of commission or omission, or wrong situational interpretation. We can only seek to improve by reducing errors in obvious circumstances and sequences of event happenings. It is extremely difficult, if not impossible, to predict all circumstances, try as we will. The human being will always find ways of circumventing even the best of systems designed to protect safety. Unfortunately, it is a fact of life.

If you or I turn a switch the wrong way, or misread an instruction for operating a machine, or fail to signal when turning a corner or fail to stop at a stop sign, we are making errors. They may or may not lead to harmful results or an accident—electrocuting oneself, reaching into moving lawn mower blades, hitting another vehicle. Perhaps they are only part of the cause: the switch was wired to the wrong wire, the mower should have stopped but did not, and the other vehicle was traveling too fast. Thus, we have multiple contributory causes.

So-called "root cause" analysis is used to estimate the causes of an error or accident and to categorize the combination of causes. But to assign a degree is difficult, since it is actually embedded in all the actions taken and may be from many causes such as inattention, ignorance, mistake, willful act, misunderstanding, or simply lack of knowledge.

Human error is usually stated to be well over 50% of the "cause," however that is defined, and can be nearly 100% of the total "blame" or cause. Thus, in auto accidents over 80% is about the fraction attributed to "driver error"; in aircraft accidents at least 60% too is assigned to "pilot error"; in safety analyses for nuclear power plants, perhaps 30% is directly due to "human error." We would argue that it is all really due to human error, depending on how the causes are classified.

Such error contributions cannot be separated from other causes easily and are integrated or embedded in the safety statistics and reporting. Even in reporting systems that categorize according to cause, the allocation attributed to human error is subjective. *But the human is involved in almost all steps and parts.* The whole technical discipline called "human factors" is devoted to trying to understand and disentangle the factors that explain and contribute to human error in the total operation and to design systems (all the way from training to the design of the human–machine interface) that are less prone to error.

Human error is the base cause, the irreducible initiator and catalyst of all major accidents and most minor ones. We find that humans—who are good at learning—follow a Universal Learning Curve, and adapt and learn well from their mistakes to reduce errors. This is almost Darwinian error theory—only those who learn from their mistakes and their errors survive, or at least survive better than those less able to learn.

But nothing is error-free, so what can we do? Can we make technological systems that are error-tolerant, so that no matter how badly we humans may screw up, the consequences (in death, fear, or injury) are acceptable?

The occurrence of error in human actions, thoughts, and deeds has been the subject of intense study. Like all psychometrics and psychology, it is largely empirical and observational and seeks to link the way humans function in making decisions and taking actions, by observing the rate at which mistakes occur—of whatever nature or cause. But this work cannot predict accident rates or provide numerical guidance about the error rate. Simple "static" models based on observed actions place stress factors on the situational assessment, so errors are more likely when the stress and distraction levels are high. If the minimum error rate is about 1%, then that says a human is likely to make an error no less or no more than about 1 time in 100. But not all such errors give accidents or consequences, and they often need to be in combination with other failures or errors. Thus, complex event trees are constructed to track the likely consequences, and the causes become complicated by all the different "contributors" to the cause. But the error rate deduced is only as good as the database used, and the dynamics of learning and time dependence are only crudely modeled.

The emergence of latent errors can only be observed and quantified by observation, since they cannot be determined by testing. Because we observe the consequences of errors each and every day, manifested as auto accidents, aircraft crashes, industrial injuries, and medical mistakes, we asked the simple question: what do these data tell us? What error rate is embedded or latent in the data? Is there a common error rate behind all these incidents, events, and injuries, as the consequences of the errors are called?

## WHO IS TO BLAME?

We assign and partition blame, just as the courts do when assigning the costs and recompense for injury or death.

The thorny question of liability arises. If there has been an error with an actual or damaging consequence, then injury has been done, and someone is to blame. But to admit to an error is to admit to liability, fiscal, criminal, or negligent. This can have severe penalties and fines, in addition to other disciplinary actions.

Thus, reporting and admitting errors is hard to do: it is potentially self-destructive. Confidential reporting systems have been devised, but even now doctors and physicians, for example, are shy (very shy) of discussing this question. The "ambulance chasers" are never far away. It is difficult to admit of the possibility of misdiagnosis of say, the health of a patient, or the reliability of a space shuttle solid booster rocket seal, or the adequacy of a Three Mile Island emergency cooling system. But each, at its own level, is an error—equally important and perhaps equally likely.

But we, as a technologically informed society, do not treat all errors as equal. We only tolerate errors that we think are unavoidable, and treat other errors as punishable or reprehensible. Yet errors and mistakes are not—by definition—made on purpose.

When we board a plane, drive our car, visit our doctor, take a pill, or turn on a light, we assume we know what we are doing. Yet we are now exposed to errors—the errors that others will cause, whose consequences will affect us; and the errors we will make ourselves, thus exposing other humans to risk.

## MANAGING SAFELY

Can we manage the risk we are exposed to? Can managers of industrial plants, systems, and people—charged with the responsibility to manage safely—achieve safety? How should safety goals and methods be set and implemented? How can we track progress in safety? Can we understand errors well enough to be sure that we are truly safe? *Are you safe*? That is the very purpose of this text, to examine and answer that last question.

Several years ago we set out on a journey to answer that weighty question, "How safe are you?" We were spurred on by the fact that there seemed to be discernible and universal trends in the accident data. On the way we found answers to other questions, too: on liability; on the limits on human error; on the role of active and passive errors; and on how we may manage safety.

We started first with commercial airline and nuclear plant safety data, and quickly found that all transportation, then all industry, and finally all human endeavors were sources of error, experience, and insights. They are all of real relevance to each of us. The facts we collected—and we were "data driven"—impacted on far-flung matters such as personal and professional liability, risk and insurability, the fallibility of medical practice, the safety of engineered systems, management methods and approaches, risks versus benefits, and last but not least, the safety of ourselves.

We were aided in our quest and search for the answer by many whose lifelong work is devoted to accident and error analysis, collection, and review. They unselfishly gave of their time and effort. All industries and areas we asked and talked to—save only the medical profession—were forthcoming and open about their errors, what they were, and what was recorded.

We think we now know the answer to the question "How safe are you?" and would like to share some of what we know and have discovered with our fellow humans. Any errors we may have made are ours and ours alone, and are because we are always prone to human error.

# 1

# MEASURING SAFETY

*"We have nothing to fear but Fear itself."*

—Franklin Roosevelt

## 1.1  SAFETY IN NUMBERS: THE STATISTICS AND MEASURES OF DEATH AND RISK

Recall that, in the Preface, we said the rate at which events and errors are decreasing depends directly on the number that is occurring. So we need to measure the number of events. We need to estimate our risk using the available historical data for events.

As early as 1662, John Graunt collected data from the "Bills of Mortality" for the City of London. He observed and calculated what proportion of people die from what cause, including those who were "kil'd by several accidents." This apparently innocent subject was rapidly turned to gain by Edmund Halley in 1693, the scientist and colleague of Isaac Newton, who derived the first life insurance tables. So the actuarial profession was born, and predicting the chance of an accident and the likelihood of death became a "science." Since the public was exposed to risks all the time, the actuaries collected the data and analyzed it to see and measure what the risk of death or injury was, depending on an individual's age, occupation, sex, income, and social status, among many factors.

Since that early time, and particularly in the later part of the 20th century, many of the accident and injury data have been collected by government and industry safety and regulatory bodies. They are categorized and reported as events, incidents, accidents, injuries, and the like. They are the statistics of error, and whole organizations record and analyze them, in great detail, with considerable expertise.

Thus the data are collected, catalogued and reported, archived and segregated by discipline, or government department, or regulatory body, or industry. These data are widely analyzed by insurers, regulators, and safety specialists, almost always within their own discipline or area of interest.

Accidents happen to other people: that is, until they happen to you. We do not expect to have an accident, to make an error, but it seems we do it all the time and ignore it, mostly because it is inconsequential. But when someone is hurt, or property damaged, or a life lost, or a technology imperiled, there is a larger cost, both social and personal. We expect others to be to blame, or be the cause, and we do not expect to be at risk. Even those in "dangerous" professions, such as builders, high-rise window cleaners, underground miners, forest firefighters, and mountain guides, soon become used to the risk exposure.

When we fly an aircraft, or travel in a train, we do not expect it to crash. So when a train crashed headlong into another outside London's Paddington Station, passing through a red danger signal on the way, that was unexpected. When three workers in a JCO plant in Japan mixed radioactive chemicals together in too big a bucket and made a larger than expected radioactive source, that was not planned to happen.

The *Titanic* was expected to be "super safe" with every precaution in the design against sinking, so safe that safety devices (lifeboats) were not really needed for everybody.

## 1.2  EXPERIENCE COUNTS: ACCUMULATING EXPERIENCE

Assuming that learning is occurring, and that errors occur, we have an average error rate. This is the instantaneous rate, usually the rate at which errors occur in a fixed interval of time. Since governments work in fiscal years, that is often the time taken, so we have E events or errors or accidents or fatalities per year, where E is any number.

For instance, if we are flying in an aircraft, then what matters is the flying time. If we are in an operating reactor, it is the operating time, not the calendar time. If 1000 people are working at a facility or factory, or a production line, then experience is being accumulated at a rate of 1000 working days for every calendar day. If 100 planes, or reactors, or autos, or factories operate for a day, then the rate of accumulated experience depends on the sum of all the technological experience. This is true for the next operating day and the next. This key observation was made by Ott and Marchaterre in their 1981 study of nuclear-related accidents, where they showed a downward trend of incident rate with increasing numbers of events.

Since the rate of learning is proportional to the experience, and if we assume an average rate of errors and learning over all the experience, then the rate of errors falls off exponentially with time (see Appendix A: Universal Learning Curves and the Exponential Model for Errors, Incidents, and Accidents). If we are not learning, then we have *insufficient learning,* and we may not follow a well-behaved trend of a decreasing rate of errors.

*This is a key observation: by tracking our error rates we can determine whether learning is occurring, and whether the rate of learning (observed as a decreased rate of errors) is sufficient.*

Thus, we show that the existing data do indeed exhibit trends consistent with the existence of a minimum error rate, and follow classical failure rate theory. In nondimensional form, the data follow a universal equation of form $E^* = \exp(-N^*)$, where $E^*$ is a nondimensional error rate and $N^*$ a nondimensional measure of the accumulated experience with the technological system. This equation form is also consistent with the hypothesis of the existence of a minimum attainable error. We discuss the wider implications of this hypothesis for diverse human endeavors in transportation, reactor safety, software, medical procedures, and industrial manufacturing.

But if you have an accident and are hurt, you become a number. We measure safety by the fatal accident rate, or the injury rate, being the numbers of injuries for a given number of people doing the same thing or job, for a particular length of time.

The event rate, R, say, is then the number of incidents (errors) in a given time scale divided by the amount of experience or risk exposure in the same given time scale, or symbolically:

$$R = (\text{Number of events (errors), E})/(\text{Amount of experience interval, T})$$
$$= E/T$$

How then do we measure E, the number or errors, and T, the amount of experience? Usually we make T a measure of time, so in a given year, or month or hour, we know what the rate is in that interval *of our lives*. But experience can and must be expressed in many ways. It can be measured in units such as hours worked, number of flights made, the number of train journeys, the passenger miles, the operating time, the total number of trips, the mileage covered, and the traffic density. This is a measure of the operating experience of the technological system we are interfacing with, using, or operating.

Most data sets, files, or government agencies concerned with data and safety reporting (e.g., OSHA, StatsCan, BTS, Airsafe) record the amount of experience as the elapsed calendar time, in years passing, or hours worked, or days active, or number of hours flying, etc.

We simply note that the measure of experience must be a meaningful one, openly recorded, and, one hopes, consistent with other data collection in other areas. We call this measure the "accumulated experience." It is distinct from the elapsed or calendar time, which is a geophysical or arbitrary human measure and corresponds more to the data collection interval or method, or to the annual budget renewal or allocation process.

We have found, moreover, that the accumulated experience (accN, we shall call it for short) arises naturally from the theory of errors and must be related to or be a measure of the risk exposure. Thus, measuring aborted landings against hours flown is not correct, as it is a short time in the landing phase. All data must be analyzed in terms of the appropriate accumulated experience, and only the accumulated experience.

Occasionally, the experience gathered over years is invariant, because the amounts are similar or constant as time passes. Only for and in this special case does the accN correspond to calendar time (years passing).

## Counting the Numbers: Event Data Recorded as Time History

We are interested in determining and quantifying the contribution of errors to accidents due to human interaction with technological machines during operation, maintenance, and design. This is the subject of "almost completely safe" systems (after Amalberti's descriptor) and so-called "normal accidents" (after Perrow's descriptor[†]). We examine the known data on errors and accidents to see whether we are indeed learning from our mistakes, and whether and how regulatory safety management and event reporting systems are effective in reducing error and accidents. We need to determine an answer to the question "How safe are you?"—to determine whether technology is indeed becoming safer as a result of training and error reporting, and because we are taking so-called corrective actions.

Since you cannot be completely safe except by not doing anything, partaking in any activity involves a risk. Even just being alive and sitting still constitutes a risk: life is not certain to start with, so we are never without risk. As engineers and scientists, we prefer to *measure* the risk exposure and the chance of an error. Many others would prefer there be absolutely no risk, or that the risk be reduced so much that it does not matter. The risk is a *perceived* risk—more a feeling of the danger or the situation.

Being technical people, we prefer to have data, and we believe that only by analyzing data can we measure, manage, and project the risk.

Extremely useful data on events are available from many outstanding and quality sources. Examples are the U.S. Department of Transportation's Federal Aviation Administration (FAA) and the Bureau of Transportation Safety (BTS), the Canadian Transportation Safety Board (TSB), the U.K. Civil Aviation Authority (CAA), the U.S. Occupational Safety and Health Administration (OSHA), the U.S. Nuclear Energy Institute (NEI), the American Association of Railroads (AAR), the Canadian government's Industry Canada (IndCan), and the U.K. Department of Environment, Transport and the Regions (DETR). There are data from private or commercial sources, and from Europe (the EU) and other countries.

The data files can often, but not always, be retrieved in numerical or spreadsheet format over the Internet, as for example at http://www.airsafe.com. Information on individual

---

[†]Selected references are listed in Appendix C under the chapter heading in which they appear.

events or accidents is often available at media sites (e.g., www.CNN.com). Compendia of accidents also exist because we humans are obsessed with tragedies—examples of such compendia are comprehensive reference books such as Denham's *World Directory of Airliner Crashes* and Berman's *Encyclopaedia of American Shipwrecks.*

When plotted against time passing, the observed events correspond to errors and form so-called discrete-time event history data sets, with instantaneous *event or hazard rates.*

There are many ways to analyze, plot, or estimate so-called event history data, and it is important to avoid the problems of censoring and time-varying explanatory variables. As noted by Allison in 1984, it is extremely difficult to draw inferences about the effect of time when a "learning curve" effect is in existence, and establishing whether a true increase is incurred in the hazard rate.

*By using the accumulated experience as the explanatory variable, rather than simply adopting calendar time, we can naturally incorporate the influence of learning.* There are other compelling reasons for *not* using the convention of calendar time as the elapsed variable when analyzing and reporting event data. The technological system does not operate to calendar time; also, the accumulated experience may be totally independent of time (i.e., how much flying and learning is occurring as the risk exposure depends on the number of flights and planes).

The number of events is dependent on how these are defined and reported. They correspond to real-time data, with censoring possible if accidents or errors occur that are not recorded, or events simply do not occur in a reporting interval. These omissions can be by chance, since we do not know when an error, incident, event, or accident will occur.

So safety agencies, governments, and operators simply count up the errors, accidents, and events as they occur, and report them periodically as tables of accident and event "statistics," classified in some way, just like John Graunt's 17th-century "Bills of Mortality." In modern parlance these series are known as "time history event data" and record what happened, when, and how often.

Henceforth, we call the reported statistics simply "errors," equating errors as contributing to the accident, event, or incident. Thus in any interval, say a day, there is a chance of a plane crash. The chance is the same for all days, and a crash may or may not occur. We must collect data over a large enough interval to have a crash—in fact, we must have sufficient crashes to make a reasonably accurate count. For rare events this is a problem, because there are not many such errors. Lack of data gives rise to surrogate measures of safety performance, since crashes are actually rare events. These surrogates are assumed to be measures by which "safety" in the nonfatal sense can be measured and reported.

Thus, we have to define "significant events," such as loss function of equipment, engine failure, sticking safety mechanisms, failure of a safety system to actuate, a near-miss, or failure of an electrical system. The more complicated the system, the more of these potential failures exist, and their failure is covered by detailed safety analyses, risk assessments, or fault trees, failure mode and effects analysis, or deterministic safety analysis. Specific safety targets are invoked. The more important systems derived from such analyses are the ones that are tracked, monitored, and maintained most carefully.

The data are often required to be reported to government and regulatory agencies charged with monitoring the "safety" of various activities, such as driving road vehicles, operating mining gear, operating an aircraft, or maintaining a nuclear reactor. Deviations from expected behavior or required practice are recorded and reported. If an injury or a significant failure of a system occurs, then additional diagnostics must be done on the "root cause." The numbers of incidents, reports, errors, or events are recorded often as part of safety indices, where the number is tracked and goals set for decreasing the number of events. Performance is measured against such indices or trends, and design maintenance or procedural changes may be required. The cost of such changes and the benefits are often weighed against the apparent change in risk.

*By all the definitions of this traditional safety reporting, safety is improved if these safety measures or indices of event, incident, and accident rates are decreasing with time.*

More important than the numbers themselves, each recording its individual significant failure or tragedy, the key is what we learn from the data, how we analyze and use the information. If that is just to assign cause and blame, that is not the full use. For example, fewer auto accidents per mile traveled or per year would mean fewer deaths and greater safety. But as the number of vehicles increases, the accident rate may increase; so additional safety measures (devices, rules, training, or protections) have to be implemented to correct the adverse trend.

## 1.3 ERROR AND ACCIDENT RATES: THE INSTANTANEOUS, ACCUMULATED, AND CONSTANT RATES

We now define three measures of error rates that we found and recommend as useful for analyzing event history data. They provide the basic information needed for tracking, interpreting, and trending. Our basic method—we call it the Duffey–Saull Method or DSM for short—relies on comparing the various observed rates, trends, and the expectations.

The usual way of analyzing is to count the number of events (accidents, injuries, or incidents) in a given time interval (a year, a month, or an hour). We may then calculate and plot the event data as instantaneous rates versus the elapsed calendar time. Instead, we prefer to also use the experience, N, so in that time interval we have the

Instantaneous Rate (IR) defined as:

$$IR = \text{(Total number of events in a given interval of experience, E)}/$$
$$\text{(Experience in that same interval, N)}$$
$$= E/N$$

This is then a measure of the rate at which errors are occurring or have occurred at any given instant. These rates are usually and for conventional convenience recorded as events per unit time (events per year or hour) or as a measure of the experience (events per mile, per flight, per working hour). This is usually referred to as the *hazard rate*.

Now experience has already been accumulated up to that time. We have operated the equipment, plants, or machines for some time before. The Accumulated Rate (AR) is defined as:

$$AR = \text{(Total number of events in a given interval of experience, E)}/$$
$$\text{(The sum of the total experience accumulated up to that interval, accN)}$$
$$= E/accN$$

This is a measure of the rate at which errors are occurring over all experience, up to and including the latest interval. This is the *accumulated hazard rate*.

We have been generally faced with two types of data to analyze, which need some discussion. The first (type A, say) is when we have found a time series where we have technology accumulating experience, for a varying sample size or population, being *events that have a given historically accumulated experience* (e.g., records for the number of injuries in mining for the workers in each and every one of the last 30 to 100 years). It is then easy to calculate and plot the trends of the IR and the AR versus the measure chosen for the accumulated experience for that given sample.

The second (type B, say) is when we have found a snapshot of data where we have technology accumulating differing experiences, for varying sample sizes or populations, being *events that have many different accumulated experiences* (say, data for the number of industrial accidents in a given year for many different countries or states for their varied working populations). In this case, it is necessary to calculate and plot the trends of the IR and AR for the chosen measure for the different accumulated experiences for those different samples.

When the Instantaneous Rate IR or the Accumulated Rate AR is plotted versus the accumulated experience, accN, we find the typical trends shown in Figure 1.1. Here the points are the actual automobile fatal injury rates for the 50 U.S. states for the accumulated experience of the past ~10 years, with the scatter being the variation between states.

This plot or data could lead to a discussion of whether one place is safer to drive in than another. As we show later, the answer is "yes": it depends on the traffic density, since driving in congested busy streets increases the collision risk.

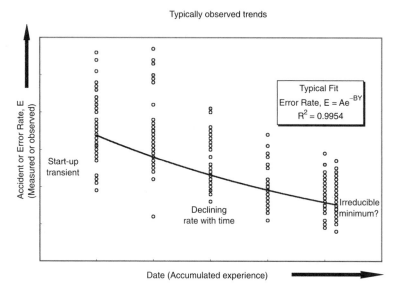

*Figure 1.1 An example of a possible learning curve: traffic fatalities over the past 10 years in 50 U.S. states showing a declining trend. (Data source: U.S. BTS, 1999.)*

This AR and/or IR graph is a plot where the experience is plotted along the horizontal axis and the event rate along the vertical axis. Since there is a lot of scatter, a line can be drawn through the data (called a "best fit" line) that gives the average and the variation of the rates. A typical fit is the exponential shown, being a line of decreasing slope and magnitude. The rate declines with time, or increasing experience, and tends towards an apparently irreducible minimum below which it cannot go. Changes in technology, or vehicle type, or local weather are all smeared (averaged) over in this plot. The presence or absence of safety devices or features such as air bags, anti-lock brakes, speed limits, multilane highways, or commuter lanes are all of second order and included in the overall data.

Overall, the death rate is just falling. If we did nothing, added no new safety devices, road improvements, or collision damage control, the rate would still fall as we learned, and there would be only a second-order effect. If alcohol consumption by other drivers increased overnight, or roads fell into sudden disrepair, or cars became crashproof, or fewer cars were on the road, then we would see a change in rate other than a steady decline. These are the *uncontrolled external factors*, over which we as drivers have no control, and would compound our chance of an error.

Basically the IR can be constant, fall, or just wander around, fluctuating as events occur or not. However, the AR usually indicates a declining trend becoming insensitive to occasional events as experience is accumulated. This is like tracking your fuel economy over more and more miles or kilometers. Over short intervals the instantaneous value (miles per gallon or km per liter) fluctuates a lot but becomes more constant or gives a more invariant and sounder average value as more mileage is accumulated. The accumulated rate is more and more insensitive to the instantaneous fluctuating rates with a regression toward the minimum, *not* to the mean. So in the same way that we may stop and start, or not, an error may occur in an experience interval, or it may not. We cannot say when it will occur; there is just a chance or probability at any time that it will.

This apparently declining hazard rate is partly due to having a nearly constant hazard (see the discussion by Allison, 1984). Thus, the rate for comparison purposes is given by the constant rate, CR, which might have been expected if there were no changes from the hazard or error rate CR(0) from when we started. Assuming we started at some initial experience N(0), and now have an experience accN, the CR(N) is then given by the change in relative experience:

$$CR(N) = (\text{Initial Observed Error rate, } CR(0)) \times (N(0)/accN)$$

This is a measure of the rate if it is invariant with accumulated or increasing experience and falls to zero as infinite experience accumulates. In practice we always have a finite amount of experience, and records and data are only available for selected or recorded times. We may term this the *expected rate* since our expectation is that things will not change or at least should be better. By tracking the CR, either as a historical or target value, we can see how we are doing, or how well we are learning or forgetting. The CR is a measure of what we might have expected to happen if we did nothing but retain a constant error rate.

It turns out, as we will demonstrate using many such data of incidents, events, deaths, and injuries, that the IR, AR, and CR are very convenient measures and ways to track safety performance. Our approach (the DSM) compares these different measures of risk to establish trends and predictions.

There is also another measure utilized in the DSM, a measure that is one of the foundations of the analysis of data. It is derived from a theory that assumes errors cannot be reduced below an apparently irreducible and certain minimum, even as we go down the learning curve. *This is the Minimum Error Rate (MER) theory*, and for those with sufficient curiosity it is described in more detail in the mathematical Appendix A. The result of the analysis, which is also discussed in more detail in a later section, is the *minimum error rate equation* (MERE), which results in the so-called exponential model for the event rate. Thus, the rate is described as a falling away function of the accumulated experience, N; that is, the error rate in functional form is just like the curve shown and drawn in Figure 1.1:

$$\text{Error Rate, } E = IR \quad \text{or} \quad AR \sim \exp(-N)$$

We will give the true or exact form later but for the moment, simply note that the MERE implies an error rate that follows a learning curve. We observe an initially high rate when we start up our system, which then is falling away as experience is accumulated, down to a finite or minimum value. Basically, errors decline with increasing experience.

Such a declining learning curve has been postulated before. The pioneering analysis by Ott and Marchaterre divided the contribution to the observed incident rate into two components. The largest was an initially declining trend due to a residue of deficiencies that somehow slipped through the design process and were called "unidentified failure modes" (UFM in their terminology). As experience was accumulated, these UFM appeared and were observed and resolved as previously unnoticed errors. Eventually, what is left was termed the "residual risk" due to the underlying "identified failure modes" (or IFM), which was recognized in the design process and subject to all the engineered safety measures. The only way to reduce the IFM further was by research and development, by operational improvement, and by design and review process enhancement.

The UFM were stated by Ott and Marchaterre to be "normally related to human errors," and their real issue was to determine what other errors were still left or "lurking in a certain technological system." As we shall see later, we now call these hidden or unrecognized errors by the name or classification of latent errors. By studying just five events, Ott and Marchaterre argued for the existence of a learning process and further asked whether studies of other technologies could provide answers to the following questions:

- How does learning reduce the error rate?
- Is the transition between the two types of errors discernible?
- What technically determines the residual (background) error rate?

By examining a much wider range of data, industries, and activities, as suggested, we can now provide answers to these questions. We can show precisely that the human error contribution is, in essence, omnipresent and such that the rate of learning is highly dependent on the accumulated experience base, whereas the attainable residual (or minimum) error rate is not. We show that this is true for a vast range of human technological activity, which implies it is universally true.

### *Knowing and Defining Risk*

We need to define what we use and mean here by "risk."

Risk is a widely used and abused term in modern technology and society. We have risks all around us: risky investments, health risks, unacceptable risks, societal risks, high-risk industries and low-risk occupations, and both risk-adverse and risk-taking behaviors. In the DSM approach, we adopt the simple classic dictionary sense, in that risk is the possibility or likelihood of a bad consequence, such as danger, injury, or loss. More specifically, *risk is the chance of an error causing you harm*, where chance is an undesigned or fortuitous occurrence or cause.

This risk is real, in the sense that it is measurable, derivable, or analyzable from actual event data and is due to human intervention somewhere in the related affairs of the universe. The risk can produce measurable consequences, or it may simply be a threat. The relative risk in DSM is conveniently measured by the size of the error interval for each separate risk category, using the IR and the AR data.

This idea is somewhat different from the risk definition used in formal and theoretical "risk assessments," such as facility safety reports and environmental impact analyses. These can be very large, thorough, and expensive. There, the effective risk, R, is defined as the product of the Probability, p, of an event happening times the Consequences, C, on a representative population sample, so that the Risk Equation is:

$$R = p \times C$$

The probability is the number of times an event might happen during the facility use— say, an earthquake or a pipe failure per operating year. The consequences can be estimated in terms of possible or potential adverse effects, such as death, injury, or disease. The overall risk is the sum total of the entire individual ones, taken over an assumed exposed population.

Relative risk is estimated by the size of R for different facilities, options, and designs, and the risk analysis process is often called "probabilistic risk or safety assessment," or PSA for short. Therefore, it is a very useful tool for comparative safety system design and assessment purposes, even though the events studied may never occur. This Risk Equation leads to high risk/low consequence cases being apparently identical or numerically equal in risk to low risk/high consequence cases. Our human risk perception does not seem to accept this apparent identity, fearing the large unknown risks more, while being more accepting of the smaller or seemingly reasonable ones even though the hazard actually can be greater.

Thus, the fundamental difference is between the theoretical or estimated risk from PSA, versus the real personal data-based risk from DSM. Ideally, these should be the same, but we may be dealing with hypothetical designs, risks, and populations in PSAs, for the purposes of licensing and safety system design. We may use PSAs to postulate many different scenarios or initiating events, ascribing their occurrence rates based on data where and if we may have it. Instead, in the DSM, we just look at the data, compare to the MERE theory, and then make a prediction based on the exponential model.

### The Minimum Error Rate Equation (MERE)

For those with a mathematical bent, the theory is given in more detail in Appendix A and results in a minimum error rate equation (MERE).

At any state or stage in experience, the rate of reduction of errors (or accidents) is proportional to the number of errors being made, A. In mathematical terms:

$$dA/d\tau \propto A$$

where $dA/d\tau$ is the rate of reduction of errors, A, with increasing accumulated experience, $\tau$, or

$$dA/d\tau = -k(A - A_M)$$

where A is the number of errors and $A_M$ the minimum achievable number, k is the learning rate characteristic (time) constant, and $\tau$ is the accumulated experience. The learning rate constant, k, is particularly interesting, since a positive value (k > 0) implies a *learning curve* of decreasing errors with increasing experience. On the other hand, a negative value (k < 0) implies *forgetting*, with increasing errors.

The resulting differential MERE is directly soluble, and the solution form corresponds to the so-called exponential model for the error rate. The errors decline to a minimum rate as experience is gained, such that we have the Universal Learning Curve (ULC) given by

$$E^* = \exp(-N^*)$$

where "exp" is the exponential function that traces out a curve, as in Figure 1.1.

$E^*$ is the nondimensional error rate, normalized to the initial, $A_0$, and minimum, $A_M$, error rates, and is written:

$$E^* = (1 - A/A_M)/(1 - A_0/A_M)$$

$N^*$ is the nondimensional accumulated experience, normalized to the maximum experience, $N_m$ or

$$N^* = N/N_m$$

For the cases or data when the initial error rate far exceeds the minimum rate ($A_0 \ggg A_M$), the simple MERE form is applicable. In this approximate form, the error rate is given by the deceptively simple and incredibly convenient exponential model:

$$A = A_M + A_0 \exp(-N^*)$$

In words, the error rate, A, at any accumulated experience, $N^*$, is equal to the sum of the minimum rate, $A_M$, which occurs at large accumulated experience, plus the rate achieved via an exponentially declining rate from the initial rate, $A_0$.

We use the above equation form repeatedly for data analysis: in fact we find all data that have learning do agree with this model and the MERE hypothesis.

We can include forgetting simplistically in the MERE theory as a nondynamic effect, shown in Appendix A, where we find that forgetting simply decreases the overall observed learning rate. The rate of error reduction is reduced such that the nondimensional error rate becomes:

$$E^* = \exp(-K^*N^*)$$

Here $K^*$ is a measure of the *ratio* of the forgetting, F, to learning rate constants, L, and is given by:

$$K^* = L(1 - F/L)$$

If the forgetting rate exceeds the learning rate, $F/L > 1$, and $K^*$ is negative so that error and event rates increase.

Omitting to do something is inherently a different matter from learning how to commit to doing it. We do not know *a priori* what the magnitude of this ratio, $K^*$, might be. The rate of forgetting could be dependent on the technological system, through training, complexity, and other influences. We will bear this possible effect in mind as we make our comparisons with actual error data.

We will show that the apparent numerical differences in risk (due to human-caused events, accidents, injuries, deaths, errors, crashes, and faults) often arise because of where the industry, operation, system, factory, or technology lies on the Universal Learning Curve. We will show, using the DSM, how a significant and common value for the minimum error interval lies hidden beneath the mass of data. The human risk perception that all these events are somehow separate and not related we will show to be false; and the search for "zero defects" will be shown to be a valuable but at the same time an apparently futile endeavor.

### Using the DSM Concepts: A Simple Example Using Traffic Accidents

We now illustrate the use of our DSM approach to help enable everyone to use it. We describe a simple example application that shows how to do the actual calculations for the error rates (the IR and the AR) using a choice for the accumulated experience basis. A similar approach works for almost any error, accident, or event data set. It is useful to have a spreadsheet PC program such as Excel available, but this is not essential, although we did all our work using it.

Here are the basic facts for a typical year, 1997, from an excellent data source, *Traffic Safety Facts 1997*, of the U.S. Department of Transportation National Highway Traffic Safety Administration:

Motor vehicle travel is the primary means of transportation in the United States, and elsewhere, providing an unprecedented degree of mobility. Yet for all its advantages, deaths and injuries resulting from motor vehicle crashes are the leading cause of death for persons of every age from 6 to 27 years old (based on 1994 data). Traffic fatalities account for more than 90 percent of transportation-related fatalities. The mission of the National Highway Traffic Safety Administration is to reduce deaths, injuries, and economic losses from motor vehicle crashes.

Fortunately, much progress has been made in reducing the number of deaths and serious injuries on our nation's highways. In 1997, the fatality rate per 100 million vehicle miles of travel remained at its historic low of 1.7, the same since 1992, as compared with 2.4 in 1987. A 69 percent safety belt use rate nationwide and a reduction in the rate of alcohol involvement in fatal crashes to 38.6 percent were significant contributions to maintaining this consistently low fatality rate. However, much remains to be done. The economic cost alone of motor vehicle crashes in 1994 was more than $150.5 billion. In 1997, 41,967 people were killed in the estimated 6,764,000 police-reported motor vehicle traffic crashes; 3,399,000 people were injured, and 4,542,000 crashes involved property damage only.

The detailed data behind this statement for 1997 are available from the NHTSA, and for 1960 through 1998. Most of us are familiar with tables of data giving, say, the number of accidents in a given year. We show this format for the traffic fatalities in a given year in Table 1.1. Reading across a row, the year is in Column A, the number of fatalities (deaths) for each year in Column B, and the number of millions of miles driven, Mm, in that same year in Column C. The data for each succeeding year appear in each succeeding but different Row.

The IR, of fatalities per million vehicle miles traveled, is derived by dividing the number of deaths by the number of miles driven in any given year, or:

$$\text{IR per Mm} = \text{Column B/Column C}$$

The result is shown in Column D as the IR, converted to per 100 Mm traveled to make it a nearly integer number as in the NHTSA reports.

We need to analyze this rate in the DSM form based on the accumulated experience. There is no reason to use a calendar year as the basis for reporting and analysis, other than for fiscal and human convenience. The number of miles traveled increases from year to year, more than doubling over the time span. *The choice made here for the accumulated experience is then the sum of the miles traveled.* This number is obtained by successively adding the miles traveled for each year, to give the data in successive rows of Column E in 100 million miles, Mm. The assumption is that we learn by actually driving, and this is the relevant measure for the accumulated experience, accMm, for this example.

The AR is obtained by dividing Column B by Column E in a given year, or:

$$\text{AR per 100 accMm} = \text{Column B/Column E}$$

*Table 1.1 Example of Data File and Calculations for a Typical DSM Application, as Taken from an Excel Spreadsheet of Original Data Downloaded from the U.S. DOT NHTSA*

| A | B | C | D | E | F |
|---|---|---|---|---|---|
| Year | Fatalities | Vehicle Miles Traveled, Mm | IR, Fatality Rate per 100 Million Mm | Accumulated Miles, 100 Mm | AR, Fatality Rate per 100 accMm |
| 1966 | 50,894 | 925,899 | 5.5 | 9,258 | 5.497 |
| 1967 | 50,724 | 964,005 | 5.3 | 18,899 | 2.684 |
| 1968 | 52,725 | 1,015,869 | 5.2 | 29,057 | 1.814 |
| 1969 | 53,543 | 1,061,791 | 5.0 | 39,675 | 1.350 |
| 1970 | 52,627 | 1,109,724 | 4.7 | 50,772 | 1.037 |
| 1971 | 52,542 | 1,178,811 | 4.5 | 62,560 | 0.840 |
| 1972 | 54,589 | 1,259,786 | 4.3 | 75,158 | 0.726 |
| 1973 | 54,052 | 1,313,110 | 4.1 | 88,289 | 0.612 |
| 1974 | 45,196 | 1,280,544 | 3.5 | 101,095 | 0.447 |
| 1975 | 44,525 | 1,327,664 | 3.4 | 114,372 | 0.389 |
| 1976 | 45,523 | 1,402,380 | 3.3 | 128,395 | 0.355 |
| 1977 | 47,878 | 1,467,027 | 3.3 | 143,066 | 0.335 |
| 1978 | 50,331 | 1,544,704 | 3.3 | 158,513 | 0.318 |
| 1979 | 51,093 | 1,529,133 | 3.3 | 173,804 | 0.294 |
| 1980 | 51,091 | 1,527,295 | 3.4 | 189,077 | 0.270 |
| 1981 | 49,301 | 1,552,803 | 3.2 | 204,605 | 0.241 |
| 1982 | 43,945 | 1,595,010 | 2.8 | 220,555 | 0.199 |
| 1983 | 42,589 | 1,652,788 | 2.6 | 237,083 | 0.180 |
| 1984 | 44,257 | 1,720,269 | 2.6 | 254,286 | 0.174 |
| 1985 | 43,825 | 1,774,179 | 2.5 | 272,027 | 0.161 |
| 1986 | 46,087 | 1,834,872 | 2.5 | 290,376 | 0.159 |
| 1987 | 46,390 | 1,921,204 | 2.4 | 309,588 | 0.150 |
| 1988 | 47,087 | 2,025,962 | 2.3 | 329,848 | 0.143 |
| 1989 | 45,582 | 2,096,456 | 2.2 | 350,812 | 0.130 |
| 1990 | 44,599 | 2,144,362 | 2.1 | 372,256 | 0.120 |
| 1991 | 41,508 | 2,172,050 | 1.9 | 393,976 | 0.105 |
| 1992 | 39,250 | 2,247,151 | 1.8 | 416,448 | 0.094 |
| 1993 | 40,150 | 2,296,378 | 1.8 | 439,412 | 0.091 |
| 1994 | 40,716 | 2,357,588 | 1.7 | 462,988 | 0.088 |
| 1995 | 41,817 | 2,422,775 | 1.7 | 487,215 | 0.086 |
| 1996 | 42,065 | 2,485,848 | 1.7 | 512,074 | 0.082 |

The answers are shown in Column F as the fatality rate per 100 million accumulated miles, accMm.

The number of fatalities in a given year has only dropped by 20%, but we can see in Column D that the *rate* (IR) plummeted from ~5.5 in 1966 to about 1.7 in 1998, a factor of more than 3. Is that due to the added safety measures such as airbags and seat belts? Ralph Nader's exhortations? Better drivers? Speed limits with safer roads and cars?

How much decrease would have happened anyway as we learned to adopt this technological system into human society? Why is the rate not decreasing much any more? Why is the rate the value it is now, and is it unusual? Will or can changes in the technology change or reduce the death rate? How do the personal hazards of driving compare to using or choosing other transport modes? What will be the death rate in the future? Are we improving continuously and as fast as we should be?

These are typical of the tough and important questions that we can examine and, we hope, gain some extra insight into using this DSM-type of data analysis.

So there we have it: Table 1.1 is now a typical DSM data set from which we can plot trends and draw new conclusions—by inspecting the numbers and by graphing the IR (Column D) and the AR (Column F) versus the experience (Column E). We will see how to do that later.

Let us now see why accidents (errors) happen, how often they happen, and what we may expect in the future.

# 2

# TRAVELING IN SAFETY

*"Be afraid . . . but not very afraid."*

—*Elle* Magazine

## 2.1 THE CRASHES OF AIRCRAFT: THE CONCORDE

### *World Aircraft Accidents: Now and in the Future*

Aircraft crashes continue to happen and to make headlines. Near the start of the new millennium, on July 24th, 2000, a technological wonder of the last millennium, the Concorde supersonic plane, took off from Paris as Air France flight 4590 and crashed in flames a few minutes later. In full view of modern media, a video tape recorder, the fatally impaired plane struggled to rise and plunged, killing all 109 souls aboard and four on the ground. It was not that such crashes are commonplace. Concorde—like the *Titanic*—was the peak of technological accomplishment at the time of its introduction, the only commercial aircraft faster than sound, and a marvel of engineering and international cooperation and commitment.

The aircraft has diverse control systems and had been subject to rigorous testing, quality assurance, and safety analysis before being certified to fly. Immediately after the crash, its Certificate of Airworthiness to fly was revoked, as if the original had been in error or something unforeseen had occurred. The cause was claimed to be a metal strip, inadvertently left on the runway by another aircraft, which ripped a tire and caused the fuel tank in the wing to be punctured by tire debris. The fuel caught fire, the engine power was reduced, and the plane crashed. The details are given in the Rapport du Bureau Enquets-Accident (BEA), Ministère de l'équipement, des Transports et du Logement, Inspection Générale de l'Aviation Civile, France, of January 2002.

Was the cause due to an inadequacy in the design, the debris on the runway, a short-coming in the safety analysis, a fault in the certification, an error in the wheel mainte-nance, poor tire performance, or the fact that the pilot did not know of the debris or the fire? Or was it a combination of these, with all these factors being contributory causes—in effect a *confluence of factors* on that day, for that plane, on that runway at that speed, with that design, on which the unexpected occurred?

Let us see what the history of other fatal airline accidents can tell us. Commercial air-craft industry data for fatal accidents are available. The system is complex with sig-nificant human (pilot) intervention. A major contributor to crashes is human error, perhaps 65% or more, and most accidents occur at times of high stress or activity, namely takeoff, approach, or landing; in reduced visibility or bad weather; or when equipment malfunctions or is somehow wrongly maintained or positioned in flight.

The inference from the worldwide *average* fatal accident rate is that it appears to exhibit no further decline over the past 15 years, being about 1 in 1 million flying hours. This is of great concern to the industry, because as air traffic grows and planes become larger, the number of fatalities will steadily rise. Can we prevent that? Do we know why it is? *How safe are we?*

Our Duffey–Saull Method (DSM) approach is data and experience driven. We use the freely available data for existing global fatal accident rates for scheduled flights from travel magazines, the Internet (www.airsafe.com), and the regulators (such as the U.K. CAA and the U.S. FAA). We do not subdivide the data or attempt to distinguish between countries, manufacturers, age of equipment, designs, and regulations. Thus, we are determining overall trends for an entire service industry, independent of detailed national or commercial differences.

In the past, tabulations and compilations of airline accident data have been given on a country or airline basis. The accidents have been extensively analyzed as to the appar-ent root causes: for example, maintenance, flight crew error, air traffic control, or mechanical failure. In addition, average accident rates, or running averages, have been examined. These methods of analysis do not really show whether there is a common underlying causality in the accident rates between, say, large and small airlines and dif-fering geographic regions.

At the website "amIgoingdown.com," entering your flight data will give you a proba-bility of crashing (all causes except hijacking and terrorism). The number is worked out using "mean variances" from apparently several years and some 63 million flights and is "weighted accordingly to produce an approximate probability." To travel from Nigeria to Canada was a chance quoted as 1 in 318,000, and from Washington, D.C., 1 in 29 million, some 100 times different. Is this right?

We plot the publicly available data on airline fatal accidents as derived from IATA, ICAO, U.S. DOT, and OAG information. The original tabulations we had were from

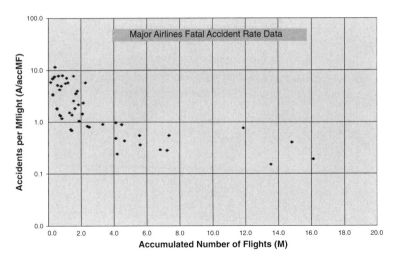

*Figure 2.1  Major fatal airline accidents worldwide for 25 years plotted as the accident rate per million flights (A/MF) versus the number of flights in millions (MF) (Data source:* **Conde Nast Travel,** *1996.)*

a travel magazine, *Conde Nast.* Data are given for individual major commercial airlines as a function of total flights for 85 airlines and contain 182 million commercial flights (MF) over some 25 years for the interval 1969–1994. This data set is therefore before the time of the Concorde crash in 2000.

What is to be used as the accumulated experience basis? It turns out that most accidents (crashes) occur during takeoff or landing, including the approach, so over 80% of accidents occur in only some 10% of the total flying time. This is when the risks and the errors are most likely: since most of the flight is usually uneventful, elapsed flying time is not a good measure of the experience, but the number of flights is.

Each airline has a different amount of accumulated experience (flights are given in accMF), ranging from some airlines with fewer than a million flights to those with 10 million or more in the United States. In effect, each airline has reached its own accumulated experience level, and we found a good set of fatal accident data in the *Conde Nast Travel* magazine. We plot these data in Figure 2.1, choosing a logarithmic scale for the accident rate in order to better illustrate and accommodate the wide range, and deleting the points for airlines with a zero rate. The horizontal axis is the accumulated experience for each airline with no fatal accidents in millions of flights, and each point is a different commercial airline; this is really an accumulated rate (AR) plot.

The trend intrigued us for some time.

The underlying trend is of a falling curve, decreasing from a high initial value toward an apparently level rate after about 6 million flights. Every airline with more than 6 million

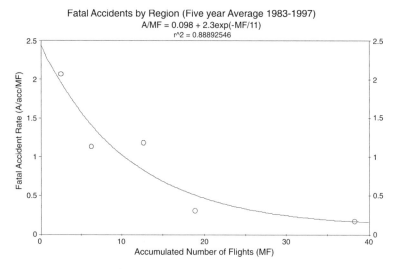

*Figure 2.2  Fatal accident rates for five world regions.*

flights has had an accident, and we suppose that those that have not yet accumulated that level of experience will eventually have a fatal accident. Recent crashes fit with that idea.

There is considerable scatter in the data (about a factor of 3 variation), and obviously there are still airlines without a fatal crash. They, and those with the lowest rates, would like to argue that they are safer. We just note that the accident rate apparently falls as experience is accumulated because there are more flights.

We found a second source of data available over the Internet (at Airsafe.com) for fatal accident rates from 1970 to about 1997, covering a similar data set of 90 airlines with ~190 MF. Moreover, these data were also grouped by five regions (United States and Canada, Latin America and Caribbean, Europe, Africa and Middle East, and Asia and Australia). The overall data trends are, of course, much the same as in the *Conde Nast Travel* magazine, so we plot just the regional data in Figure 2.2.

What we have actually done is "bin" the data or average the individual airlines over convenient blocs of countries. But is one region safer than another, in the sense of lower crash rate or chances of a crash? Is that difference real? That is something we all might like to know!

We see the same trends evident in Figure 2.2 as in Figure 2.1 (it is really the same data set, but re-binned), despite some detailed differences in the database. The exponential line drawn through the data is derived from the mathematical curve-fitting routine

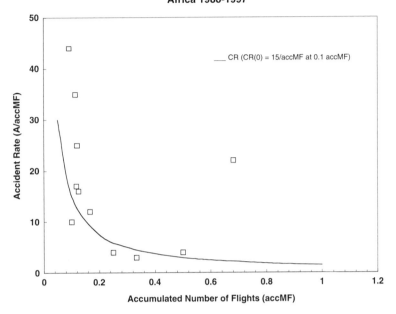

**Figure 2.3  Data for hull losses in Africa. (Data source: Russell, 1998.)**

TableCurve2D and is

$$AR \ (A/accMF) = 0.1 + 2.3 \ exp(-accMF/11)$$

This equation and curve fit have some interesting properties. The "tail" or minimum of the curve at large accumulated experience is *predicted* to be the lowest (asymptotic) value of ~0.1 per million accumulated flights, somewhat lower than the value of ~0.2/MF for the North American region grouping. The higher value is ~2.3 per million flights and is weighted by the data from the Africa region grouping, with the lowest accumulated experience. So the apparent span of values from the predicted lowest to the possible lowest is some 20 times as the experience changes. But this range does not yet mean one region is safer than another; rather, it means just that the rate seems to fall as more experience (measured in accMF) is accumulated.

A third source of data is from the airline insurance industry, but the data are not published publicly. We simply note that, as far as we know, these data yield regional and average statistics comparable to the above, with similar trends. As an example, Figure 2.3 shows Russell's (of Boeing) recent data for Africa for hull losses as opposed to fatal accidents alone, where there are more nonfatal hull (aircraft) losses than fatal accidents. We have plotted a line through the points on the graph as the constant rate, CR, as given by an initial hull loss rate of ~15/accMF at very low experience.

For Africa, there is just too much scatter and not enough accumulated experience at the higher values to make a confident prediction of the minimum value for large numbers of flights, but the overall trend is down as experience (flights) are accumulated.

All these data and curve fits are telling us something, and we make the following six observations:

1. The trends for all airlines *with an accident* show that there is an apparently declining accident *rate* (accidents per megaflights, A/accMF) with increasing number of (mega)flights (accMF).
2. The trend falls to a minimum rate that is increasingly *insensitive* to the increasing number of megaflights when the number of flights exceeds about 3 to 6 million (3MF).
3. The data trends are apparently within the data scatter largely *independent* of airline, aircraft type, country of origin, geographic region, or area of operation.
4. The plot can be readily *converted* to other measures of experience if needed to the number of takeoffs and landings (TOFLs) or the average number of flight hours assuming either (a) an average of two TOFLs per flight or (b) an average flight duration, if known, which is typically some 3 hours and 20 minutes for the United States.
5. The average trend of accident rate (A/accMF) apparently falls with increasing flights to a minimum value *just like a learning curve*, and all airlines have had an accident by the accumulation of ~6MF of experience.
6. The same overall trends occur on a regional basis as with the individual airlines (declining with increasing number of flights); moreover, individual airlines in different regions have accident rates, A/MF, comparable to those for whole regions with a similar number of total flights.

*Thus, we may conclude that there is a common trend in all these data: no one area, region, or plane type is much safer than another.*

The observed apparent asymptotic or minimum accident rate (of about 0.15–0.3 A/MF) changes from time to time as different airlines with the most accumulated experience (mainly in the United States) have accidents at different times. The higher accident rates, whether by region or individual airlines, are *always* for those with the lesser numbers of megaflights or least amount of *accumulated experience*.

### What Is the World Error (Aircraft Fatal Crash) Rate?

We are now ready to answer this question, which we illustrate by analyzing and predicting future event rates. Let us look again at the latest fatal accident rate data from the CAA for North America (NA) and the European Joint Aviation Authorities (JAA) for 1980–1999 (AirClaims, 2000), and the latest updates for the world commercial airlines for 1970–2000 available from (Airsafe.com, 2000). Thus, the data sets now cover

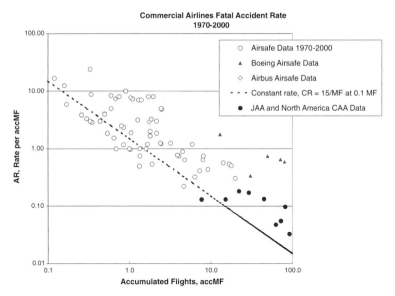

*Figure 2.4  The latest airline fatal accident rates. The CR is plotted on a log-arithmic scale: the straight line is for a constant rate of 15/MF. (Data sources: Airsafe.com and U.K. CAA, 2000.)*

more than 200 million flights, MF, for more than 110 airlines and all the latest accidents, except the Concorde. Figure 2.4 shows all the data, expressed conventionally as the accident rate per accumulated flights, A/accMF, for the airlines and aircraft types, and includes the latest data and predictions as follows:

(a) The data and the *predicted* fatal accident rates for that 1967–2000 experience; plus
(b) The data for different modern aircraft types, as given by Airsafe for Boeing (700 series) and Airbus (A300 series) commercial jets; plus
(c) The constant rate, CR, line consistent with the initial observed rate of about 15 per accMF.

The CR is designated conventionally as having a slope of minus one (–1) on such a log-log scale plot: the rate falls by a factor of 10 for each increase in experience by a factor of 10. Above about 6 accMF, the data fall above the line—so the rate is somewhat higher than the CR would predict. The data for the event rate then seem to flatten out for a greater number of flights. There is about a factor of 10 variation (scatter) in the data, and the trend is independent of airline, country, or aircraft type.

To compare prediction approaches, the apparent minimum accumulated rates for the world data using the DSM was ~0.2/accMF and for the JAA/NA about 0.13/accMF based on the 5-year average, so these are quite close. The comparisons between the AR and the CR lines and between the MERE predictions are all a little clearer on this logarithmic plot.

We tried a MERE fit of the form

$$AR = AR_M + AR_O \exp(-N^*)$$

which, for just the Airsafe data, excluding the regional grouping, is found as

AR (Accidents per accumulated million flights), A/accMF
$$= 0.2 + 15 \exp(-accMF/0.3)$$

The initial rate indicated by the DSM model is CR(0) and is therefore ~15/accMF, consistent with the straight line in Figure 2.4.

The key point is to first order, for up to ~10 accMF the accident rate is a nearly constant initial rate, CR(0), which is invariant all over the world at about 15 per million accumulated flights, 15/accMF, at least up to the maximum accumulated experience of modern airline fleets (~10accMF).

Since an average flight is some 3 hours 20 minutes ($3^1/_3$ hours), the initial error rate in elapsed pilot or aircraft flight time is 15 out of every 1,000,000 flights, with an interval of 3.3 hours. The initial interval between fatal errors (crashes) is then of the order:

~ (1,000,000 flights × 3.3 hours) divided by 15 errors
in a million accumulated flights
= one in 220,000 flying hours,

or a fatal error frequency F = 1 in $2.2 \times 10^5$ hours.

*This is a fundamental finding: the time between the confluence of events is on average the same for all commercial flying experience. We conclude that pilots all over the world try not to have fatal accidents with about the same (non) success everywhere.*

*Clearly, factors over which no individual airline, aircraft, pilot, or airspace has control have combined to produce this fatal error frequency, F, of ~1 in 200,000 hours for this technological system (travel in commercial aircraft).*

Therefore, the risk numbers given by "amIgoingdown.com" are apparently based on the usual IR statistics that fail to account for the differing accumulated experience bases embedded in the crash rates. Therefore, the crash probability numbers given—which differed by a factor of 100 for flights to Canada originating in the United States and Africa—are both misleading and incorrect. The correct value is one crash in 65,000 flights on average everywhere in the world (15/accMF) for every airline within about a factor of 10, based on the more than 20 years of published data shown in Figure 2.4, independent of the route and region traveled. The apparent differences are only because of the statistical variation with increasing experience.

## A Test of the Theory Using Concorde Data

Let us test this observation and DSM prediction with the individual and particular Concorde accident. This is a single aircraft type, flown under very special conditions, which is very different from normal subsonic aircraft. It was subject to rigorous safety review and scrutiny and was operated to the highest standards by internationally famous airlines. Up to the time of the Paris crash, the Concorde fleet had accumulated about 90,000 flights.

The accident rate, or AR, is then $(1/90,000) \times 1,000,000 = 11$ per million flights, 11/accMF. This rate is a little lower but to first order is the same as the world initial average for all airlines of 15/accMF. Hence, this terrible crash is in a sense not special but follows the world experience.

The press release on the Concorde accident available from the French Inquiry (by the Bureau Enquêtes-Accidents) states:

> During takeoff from runway 26 right at Paris Charles de Gaulle Airport, shortly before rotation, the front right tyre of the aircraft ran over a strip of metal, which had fallen from another aircraft, and was damaged. Debris was thrown against the wing structure leading to a rupture of tank 5. A major fire, fuelled by the leak, broke out almost immediately under the left wing. Problems appeared shortly afterwards on engine 2 (i.e., inner left engine) and for a brief period on engine 1 (outer left engine), because of hot gas ingestion due to fire or debris. The crew took off, then shut down engine 2, still operating at near idle power, after an engine fire alarm. They noticed that the landing gear would not retract. The aircraft flew for around a minute at a speed of 200 kt and at a radio altitude of 200 feet, but was unable to gain height or speed. After new ingestions of debris or hot gases, engine 1 lost thrust definitively, the aircraft's angle of attack and bank increased sharply. The thrust on engines 3 and 4 fell suddenly. The aircraft crashed onto a hotel.

Tire failures had occurred before on the Concorde, so we also have analyzed this failure data in more detail (see the later section on road safety) to see if a precursory indication existed related to such potentially catastrophic tire failures.

## Other Commercial Aircraft Events: Near Misses and Event Data and Analysis

Other events occur in the air: near misses, missed approaches, and go-arounds. For each of these, procedures and controls are in place to minimize the risk of accident. What do they tell us? By using the DSM, as a first step we undertook an analysis of the event data to determine (a) whether the DSM approach could be applied to real event data, and (b) whether there were any trends that could be determined.

Events occur all the time and obviously depend on how much "air traffic" is occurring. When event data are plotted versus the accumulated experience (which is now defined here as the accumulated number of flights or movements) for the CAA event report data, we find the typical trends shown Figure 2.5. Basically the IR can wander around,

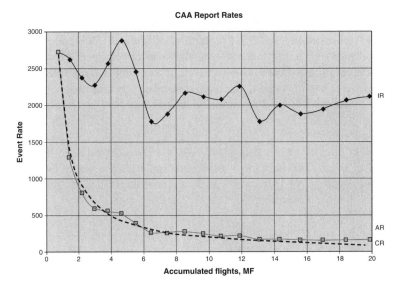

***Figure 2.5  Typical event rates: comparison of accumulated, AR, instantaneous
IR, and constant rates, CR. (Data source: U.K. CAA, 2000.)***

but the AR is indicating a declining trend. Note however, that this apparently declining hazard rate is partly due to having a nearly constant hazard (Allison, 1984). The
constant rate for comparison purposes is given by:

$$CR = \text{Initial observed error rate} \times \text{Accumulated experience ratio}$$
$$= CR(0) \times (accMF(0)/accMF)$$

In this book, we adopt universally a data-driven approach, using solely the available
operational event and error information. An event is equivalent to an error, or sequence
of errors, or a combination of acts or actions that result in an error. We avoid the inter-
minable discussion of qualitative cognitive models and of the psychological causation of
errors, and the role of separate or multiple barriers or defenses against error (Reason, 1990).
We evaluate the trends in event data and relate these to potential safety indices.

The difference is apparently startling in the trends based on the traditional IR view of
the world. It is clear that the IR is not decreasing steadily: also note that there is no evidence of asymptotic behavior, and the IR seems to be fluctuating and climbing in recent
years. It is difficult to make a prediction based on a curve fitting these IR data.

However, the AR shows an apparent steady *decline* with increasing experience, and this
line can be used to project the future of ~160 events per million *accumulated* flights or
movements. Thus we have a minimum error frequency of

$$F = (1,000,000 \text{ flights}) \times 3.3 \text{ hours per flight}/160 \text{ events}$$
$$\sim \text{one error per 20,000 accumulated flight hours, accFh}$$

So there are roughly 10 times as many incidents or events of a generally reportable nature as there are fatal accidents.

If the event rate had been constant from the start of reporting, then the *expectation* would be that the dotted *constant rate line*, CR, would have been followed. Therefore, the rate is indeed deviating and is about 60% more than would be expected from a constant rate for the experience accumulated up to the present.

The question raised is why or how the DSM for examining error rates using the AR shows declining event rates, when more traditional IR measures or approaches show an increase using event rates based solely on the current experience. The difference is clear: the AR declines steeply, whereas the traditional IR increases.

Thus, the events are following a well-behaved learning curve; the AR provides a basis to make a prediction and can be easily compared to the CR because the rate is being averaged over the entire accumulated experience.

### Near Misses: The Same Rate?

Recently, two large passenger aircraft with more than 500 people on board reportedly came within 10 meters of each other over Japan despite having visual sighting of each other for many miles. A combination of reported air traffic control and flight crew confusion, not taking correct action in response to collision warnings, caused conflicts in the flight paths. It was truly a nearly catastrophic near miss, in every sense of the word.

Is the near miss a fault of the technology? Is there a problem in procedures? Is the collision avoidance warning system at fault? Or is this another confluence of these and other factors?

It is useful to look at these factors in the sense that these errors should not happen and whether they are a precursor to a more serious accident (a fatal accident due to a collision). Such accidents have happened in the past, as in the San Diego event (September 1978). Well-defined airspace and control rules are in place to avoid the chance of a near miss. Proximity warning, ground radar, and ground controllers play a major role in ensuring spatial separation. Yet all these are violated occasionally.

To further illustrate the power and use of the approach, let us examine another independent event data set. We use the "near miss" risk-bearing event data for U.S. commercial flights for 1987–1997 Category A&B NMACs (FAA, 1999) and the Canadian Transport Safety Board Airprox data (TSB, 2000). These are loss of separation distances for aircraft in flight that represent a significant hazard potential or risk of collision, even if one did not occur because of recovery actions. The other category of near misses is basically where the separation did not or could not produce a hazard.

We need to convert these data, which usually are reported by calendar year, to the DSM format. Plainly, the accumulated experience that is relevant to near misses is the

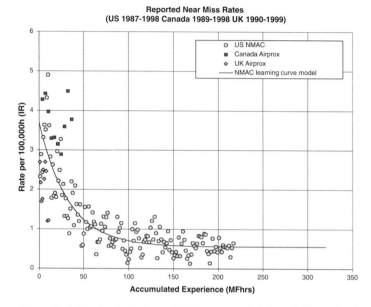

*Figure 2.6 The FAA, CAA, and TSB commercial aircraft NMAC and Airprox data and the DSM predictions to 300 million flying hours. (Data sources: CAA, FAA, and TSB, all 2000.)*

exposure in actual flight time and hence corresponds to the accumulated experience in flying hours. As further data, we also have access to the equivalent U.K. "near miss" CAA Airprox category A&B commercial data (CAA, 2000; U.K. Airprox Board, 2000). Both the U.K. and Canadian data are very much lower accumulated experience than that in the United States.

In Figure 2.6, we use the accumulated experience in millions of flight hours (MFh) as the available accumulated experience basis, and we plot the IR as conventionally given by the various reporting agencies as events per 100,000 flying hours.

These IR near-miss data follow a classic declining learning curve, whose exponential form was the statistically significant and preferred first choice. This trend is shown as the continuous line and is based on a MERE exponential model (see Equation (10) of Appendix A). From this exponential "learning curve" model for the failure (error) rate, A, the general form of the data should be

$$A = A_M + A_0 \exp(-accN)$$

Here $A_M$ is the minimum rate, $A_0$ is the initial rate, and $accN$ is the accumulated experience. So, as we might expect, the equation for the line that fits the data is:

Near misses per 100,000 Fh, $IR = 0.55 + 3.2 \exp(-accMFh/34)$

We find that the predicted asymptotic limit projected by this exponential DSM model equation is ~0.55 per 100,000 accumulated flying hours and apparently will not decrease substantially in the future. *This minimum time between errors (the error frequency, F) is 1 in 180,000 Fh (or close to ~1 in 200,000 accumulated flying hours), nearly exactly the same as the initial interval that we found for fatal accidents.*

This frequency is 10 times less than that for typical errors of a general but reportable nature. The DSM therefore allows us to compare these very disparate near-miss data sets by using as the common basis the accumulated experience.

The comparisons shown in Figure 2.6 clearly illustrate that the event rates are broadly comparable, even though the U.K. and Canadian accumulated experience is comparatively small. The DSM has brought together these three sets and shows the presence of a learning curve.

More importantly, Figure 2.6 illustrates the trend that any country or airline could be expected to follow in the future as experience is accumulated. Thus, we find the rates are comparable for equivalent accumulated experiences, across these three countries, which themselves should be following a learning curve (in effect, a mini-learning curve with limited data within a macro-learning curve of more data). This is illustrated very clearly in the AR plot in Figure 2.7, where even the limited-accumulated-experience U.K. data set can be seen to follow a trend, the dotted line being the initial constant rate, CR, expected and observed from the larger U.S. data set.

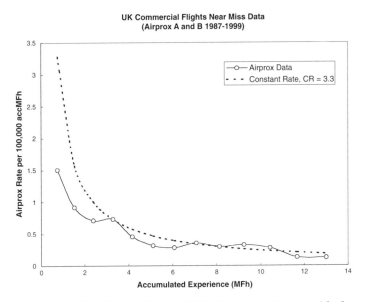

*Figure 2.7 The AR plot for the U.K. Airprox mini-curve with the predicted CR. (Data source: CAA, 2000.)*

In other words, the U.K. near misses have set off on exactly the same learning curve as the United States has taken, with an initial learning curve all of their own.

### Hazards and Risks in Other Types of Flying: Insufficient Learning

In addition to commercial airlines that carry passengers, there are private planes and charters (so-called general aviation), helicopters (so-called rotors), and short-haul limited space passenger planes (the air taxis and commuters). Crashes of private planes and charters have claimed many famous names, since the rich and the privileged as well as ordinary people often travel by private or chartered transport.

The unfortunate crash of John F. Kennedy, Jr., over the sea in July 2000 is one notable example where a relatively inexperienced pilot and his passengers died. Pilot error was inferred to be the cause as there were no mechanical problems reported or found, and the Kennedy family reportedly paid several million dollars as part of a settlement with the families of those who died.

The accumulated experience base with each type of air transport is different, but the U.S. BTS and the FAA have the greatest set of data files on incidents publicly available for 1982 through 1997. So we plotted these incident rates, IR, converting the accumulated experience to the millions of flight hours for each, with the result shown in Figure 2.8.

Immediately, we see that the "large carriers," being the commercial airlines, commuters, and air taxis, have a falling rate (as we would expect for a learning curve), but

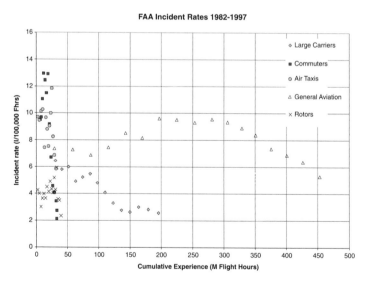

*Figure 2.8 Comparative incident rates, IR, for different carrier types. (Terminology and data source: FAA, 1999.)*

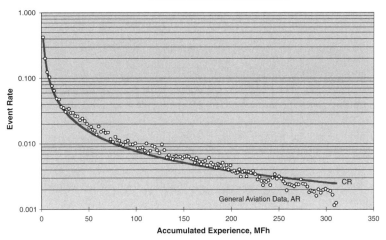

*Figure 2.9 The trend in U.S. general aviation events. (Data source: U.S. FAA, 1998.)*

general aviation does not, demonstrating a hump, despite having the greatest experience base. Can we discern more? Is general aviation a greater hazard now than in the past? Are there more errors?

We can examine these error (incident) data in more detail. The comparison to the constant rate, CR, of the general aviation accumulated rate, AR, is shown in Figure 2.9. Initially the data fell above the CR until at about 220 MFh they started to fall below. So there is some variation; however, it is clear that there is very little improvement until greater than 250 MFh, comparatively speaking. Certainly, general aviation is not learning that much from the accumulated experience. *We call this error trend, where the rate tracks a nearly constant rate or exceeds it, "insufficient learning."*

We will find other fields and examples where insufficient learning is the case: where apparently the error rate (as measured by the accident, incident, event, hazard, or reportable occurrences) is not decreasing as one might expect from a learning curve, and the errors can even be increasing. *By comparing to the CR and discerning the learning rate, the effect and effectiveness of management strategies to reduce errors and risks can be estimated.*

### Runway Incursions

A recent example and "hot topic" is runway incursions, an example in which other factors than simply one machine and one pilot are involved. The error is allowing an unauthorized intrusion of some other aircraft or equipment onto the runway when it should not be there or other associated errors, thus endangering an aircraft that is maneuvering, landing, or taking off. The rules governing runway protocol are clear, but it is also essential for the human element to understand and obey the rules, since there are no physical separation barriers, and runways are open surfaces.

The Singapore Airlines crash on takeoff at Taipei Airport is another example of the *confluence of factors*. The error (crash) occurred in poor visibility, under worsening storm conditions, with the aircraft taking off on the wrong but parallel runway, one that was closed for repair, and plowing into the concrete barriers and equipment innocently and routinely left there by the repair crews. This was a "runway incursion" of the reverse and worst kind: the plane should not even have been there!

In Milan, Italy, in late 2001 a large jet airliner collided on the ground with a small corporate jet aircraft that had made a wrong turn, in fog, causing the death of all the passengers. Reportedly, the airport had no operating ground radar at the time. The Prime Minister of Italy said it was "inconceivable that the airport of one of the most important cities in Europe is even touched by the suspicion of neglect, omission or worse."

The U.S. data on runway incursions have shown a rate increase on a per-movement basis, as shown in Figure 2.10. It is somewhat clearer on a logarithmic axis plot, since the current AR is some two times the expectation or CR value. Thus, although a learning curve has been followed, the *rate of learning* (decreasing slope) has not been sufficient. This was noted by the FAA in making runway incursions one of the agency's

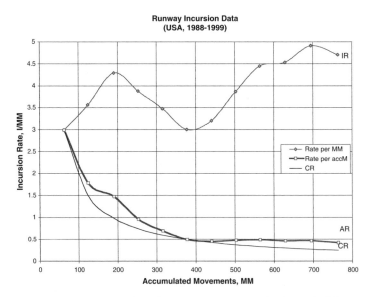

*Figure 2.10  IR, AR and CR for runway incursions in the United States for the 12 years from 1988 to 1999. (Data source: FAA, 2000.)*

top priority safety items. Having inaugurated a special program on the topic, FAA Administrator Jane Garvey said in 2000:

> I am concerned about the number of runway incursions because according to the National Transportation Safety Board and FAA data, runway incursions continue to increase. There has been a 73 percent increase in the number of reported incidents from 1993 through last year. There were 186 reported in 1993. In 1998, that number was 325. The rate increased from 0.30 per 100,000 airport operations to 0.52 incursions per 100,000 airport operations. We must reduce that rate.

What do the data indicate? What should be the management target rate, which is taken for the reduction?

The number of movements is relatively constant from year to year at ~65 million, but the rate and number of incursions have both increased ~75%. The AR per million movements (accMM) *should be* 0.24 on a learning curve basis instead of the observed 0.42, such that the *expected* rate is in fact:

$$\text{Expected rate per year, CR} = (\text{Constant rate per accMM}) \times (\text{accMM})$$

This expression gives about 180 per year, which is 56% or nearly a factor of 2 lower than is actually being observed. *So a reasonable management goal might well be a factor of 2 reductions in both the IR and AR.*

We have spent a lot of time on commercial airplanes, because that industry is a good illustration of the use of the DSM approach: it has shown and shows the value of data collection, and it enables us to check our capability to think through what the trends may mean.

We now turn to other modes of transport, to see what is happening, what the risks and errors are, and whether claims of one mode being safer than another are justified—and if so, why that might be. After all, if we can travel more safely, perhaps we should consider our choices for travel, if indeed the risk is voluntary.

## 2.2 THE SINKING OF SHIPS: *TITANIC* AND TODAY

Setting aside losses due to warfare and acts of God, there is a rich history of ships lost, foundered, and sunk available to study and marvel at. Over the centuries—indeed, since humans first took to the sea—ships have sunk as a result of collisions, groundings, fires, hurricanes, and simple leaks. The encyclopedic book *Shipwrecks* by David Ritchie lists and discusses about 1000 so-called significant (famous and interesting) wrecks, selected from some thousands of ships over the 500 years since 1496, ignoring losses in battle. Over half of these losses are recorded in the 200 years of the 19th and 20th centuries, presumably because of better record keeping and more shipping. More extensive listings that make gripping reading can be found in Hocking's two-volume *Dictionary of Disasters at Sea during the Age of Steam, 1824–1962*, as compiled and published by Lloyd's Register of Shipping. The tabulations in Berman's book (*Encyclopaedia of American Shipwrecks*) contain more than 10,000 shipwrecks for the United States alone since 1800.

The historic losses vary: from whole fleets of treasure galleons in great storms to the grounding and breakup of tankers and liners on sand bars and reefs; from the loss of merchant schooners and oil tankers in bad weather and in good, to mysterious disappearances at sea and sudden sinkings.

The exact number of lives lost for each wreck depends highly on the circumstances and varies from none to hundreds or even thousands. The number saved depends on the availability and capability of life-saving gear and effort, as well as on individual and collective heroics and sheer luck. So we must really count the rate at which ships, rather than lives, are lost. Before 1741, according to Lloyd's Registry, there was no centralized source of merchant shipwreck information. Since then, many records and indexes have been collected that are manually accessible (see http://www.lr.org for a listing of these sources).

Ships are lost because of errors: in design, navigation, seaworthiness, anchorage, or cargo. *The errors occur in combination or confluence, as we shall see.*

### *The* Mary Rose, *the* Vasa, *and the* Titanic

The *Mary Rose* was one of the great warships of England's great King Henry VIII. Experienced in battle, a flagship of the fleet, and fully refitted, in 1545 this majestic sailing ship sailed out from Portsmouth Harbor to do battle with the French. But the *Mary Rose* was loaded with more than 700 men, soldiers and sailors, and the ship had been designed for only 400. Suddenly, as the ship listed in the wind, water entered through the still-open lower gunports, and the ship sank 2 km offshore, in full sight of the assembled well-wishers, drowning 660 souls.

There were many recriminations, but no official inquiry took place. "Human error" was the cause, says a plaque at the Portsmouth Museum in England, where thousands of artifacts from the tragedy are on display, including what is left of the hull. Multiple causes were assigned: top-heavy design; overloading; inadequate crew and soldier training and responsiveness to orders; leaving the guns run out with open gunports while maneuvering; the need to be at sea and in action quickly ("pressonitis" as it is called).

A list of contributory factors is thus easy to assemble. In modern terms, these include:

    (a)  The unproven design changes and maintenance errors
    (b)  Inadequate operating and emergency procedures
    (c)  Poor training and communication systems

In 1628, just 83 years later, the mightiest warship in the world, constructed over three years from a thousand irreplaceable oak trees, and bristling with 64 large cannon, set sail from Stockholm harbor. On its maiden voyage, the great vessel the *Vasa* represented the peak of warship technology of the time. Just offshore, a mere breeze hit the sails, the vessel leaned over, water rushed in through the gunports, and the *Vasa* sank.

The reasons for the *Vasa* tragedy, given in the great inquiries that followed, were unknowingly and understandably all designed to shift the blame and assign the cause to the errors of others.

"Only God knows," said deGroot, the Shipbuilder, having built the ship to the royally approved dimensions or design requirements. "The ship was too unsteady," answered Captain Hansson, blaming a design fault. "The ship was top-heavy," said the crew, denying that any of the guns or equipment were not properly tied down. "Imprudence and negligence," wrote the Swedish King, seeking to find and punish the guilty.

We now know there were multiple and interrelated causes, which sound very familiar to us. There was what we call a *confluence of factors*, all including a human element and a whole variety of latent errors.

The Admiral failed to prevent the *Vasa* from sailing after tests on tilting and ballast showed marginal stability; the King personally approved the vessel's dimensions even though he was not a designer, and pressed for a tight completion schedule; the Shipbuilders extrapolated and scaled up the design in size and weight beyond the range of existing engineering knowledge; the Captain sailed with insufficient crew training and knowledge for safe maneuvering of the ship and with open gunports.

In modern terminology again, we have the following *root causes*:

(a) Inadequate training
(b) Poor operating and emergency procedures
(c) Faulty engineering and design quality assurance
(d) Incomplete research and development
(e) Insufficient validation
(f) Faulty commissioning
(g) Nonadherence to procedures

Could these have been foreseen? Were the errors preventable given the technology and processes of the day? Today in the Vasa Museet you can tour this great ship, painstakingly found and lifted, and a monument to modern-day salvage and restoration technology. We marvel at the detailed workmanship, the supreme craft of shipbuilding, the perfection of the carvings, and the apparent overwhelming stupidity of all those extremely able people. The errors were in the human–machine interface (the so-called HMI) with the technology.

Sinkings and wrecks at sea have now occurred for centuries, where the errors are multiple and complex with significant human (pilot/navigator/crew) intervention. A famous example is the liner *Titanic*: in 1912 the unthinkable happened and the "unsinkable" sank on its maiden voyage. The disaster showed that "man's greatest efforts in the fields of engineering and technology could be brought to nothing . . ." (McCluskie *et al.*, 1999). More than 1500 passengers and crew were lost when the gargantuan vessel went down.

The needless loss of life due to inadequate lifesaving capacity had occurred nearly a hundred years earlier, in the loss of the *Home* in 1837, where there were only two life preservers for the 130 passengers and crew!

The causes of the loss of the *Titanic* have been well researched. The inquiries of the British and Americans again cited a *confluence of (latent) factors*: navigation in an ice field at excessive speed where "what was a mistake in the case of *Titanic* would be negligence . . . in the future"; a fatal flaw in the design and use of part-height watertight bulkheads; inadequate regulations concerning the number of lifeboats to be carried; and possible negligence on the part of the captain. Once again, there was no unambiguous assignment of blame or responsibility, although many professional careers and reputations were ruined or changed forever.

In today's terms we have to say the loss was due to a combination or *confluence of contributory factors*, including:

(a) Inadequate safety design and defense-in-depth
(b) Violation of safe operating procedures ("pressonitis") and lack of adequate lifeboats
(c) Management failure

Weather and visibility conditions are often a contributor, but we assume—since we have had many centuries of practice—that sinkings and losses today are due to errors. Indeed, in all navies, the captain always takes responsibility for what happens to his ship, since all are under his command. But, just as with airline pilots, we should not place blame for every error on the captains, even though they may carry the ultimate management responsibility. They are interfacing with a technological system, which has training, navigation, and safety features embedded in the process. They are themselves on board!

### Insuring the Seas: Shipping Loss Rates from History

That said, someone has to bear the responsibility and pay for all the damages and losses, both physical and financial. In the tradition of John Graunt, this is the role of marine insurers, who must estimate the likelihood of loss and the possible costs. Human error is the main cause of major losses, not faults in the ships. The actuarial profession slaves over the data, to estimate the risk in financial and liability (and exposure) terms. Insurance data for total losses for ships more than 500 metric tons have been published as annual tabulations and compilations by insurance consortia such as the

Institute of London Underwriters, which have kindly and openly made their data available to us.

We examined the trends for total losses for about 25 years, from 1972 to 1997, covering in a given year up to ~500 Mt of shipping in about 35,000 vessels. This is about 1 million ship-years of experience. We chose shipping-years (Sy) as a measure of the accumulated experience, since the number of ships at sea can vary from year to year. Thus, the shipping years (Sy) are the number of ships at sea times the number of each of the years at sea. Since these data are usually given on a year-by-year basis, we actually converted these raw data to the yearly loss rates per thousands of ship-years, the IR (L/kSy) and the AR (L/acckSy). The world shipping *accumulated experience* is measured by the total number of thousands of ship-years (kSy) afloat since 1972, for which we had the loss data.

As shown in Figure 2.11, we see the trend that as experience is accumulated, the loss rate falls off toward an *apparent* minimum or asymptotic level, which actually corresponds to the AR very nearly following an almost constant rate, CR. The CR initial value taken was ~6L/kSy. There is some evidence of a continuing slight decrease in the number of losses, based on TableCurve software curve fits. This lower *average* rate is ~0.4L/acckSy, where the total accumulated experience since 1972 is about 1 million shipping years afloat (~1MSy). Thus, the AR and CR almost track each other, where the initial or expected IR rate is between two and six total losses per 1000 ship-years (2–6/kSy), with a current minimum near 2 per thousand ship-years. At least 50% of the losses are attributed to human error, rather than to ship structural failures and causes.

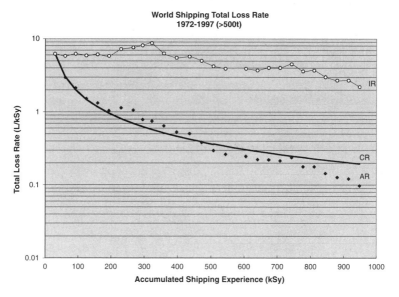

*Figure 2.11 World shipping total losses for 1972–1997 for vessels over 500 metric tons. (Data source: Institute of London Underwriters, 1999.)*

On the basis of this analysis we would expect that, with 35,000 ships afloat, losses would never be less than about

$$35,000 \times 2 \text{ per } 1000 \text{ Sy} = 35 \text{ kS} \times 2/\text{kSy}$$

or about 75 sinkings in every year into the foreseeable future (or until truly unsinkable ships are designed and in full operation!).

By valuing the cargo and assigning a usage (afloat) factor for a given charted course and speed, we can assess the risk of total loss for any given ship cargo and set the insurance premiums accordingly.

The error (total loss) frequency is today about 2 per 1000 ship years, or $2/(1000 \times 365 \times 24 \times f)$ per hour, where f is the fraction of the year spent actually at risk afloat. The answer to the multiplication is about 1 in $4,400,000 \times f$ ship-hours.

If the afloat (usage) factor f is about 90% of the year (11 months) for all the ships to spend afloat on average, with an assumed active or navigational crew of about 12, then the error (leading to total loss) frequency is of order one in 330,000 accumulated crew-hours.

This result is within a factor of 2 of our previous error frequency for aircraft (1 in 200,000 flying hours). We say this is the same "order of magnitude" (i.e., within a factor of 10). Given the data and the assumptions, this result is certainly the same order of error frequency and certainly within the error of the calculation itself.

*It would seem that a nearly constant minimum error (loss) rate has been reached in shipping* (at least for vessels above 500 metric tons), and *ships' masters and their crew try to avoid losing their ships with about the same success everywhere.* Only some significant shift in the technology can or will lower the rate, with the continued loss of shipping in good weather and bad.

### Personal Risk: Recreational Boating Deaths

The loss of life will vary when a ship sinks because the numbers of crew and passengers vary, as do the numbers of survivors plucked from harm. Thus, we cannot easily predict the number of casualties or fatalities in the sinking of large vessels.

There are also many, many recreational boats on the water, and these personal craft also have accidents. The major number of fatalities is in propeller-driven craft, with about three times as many injuries as there are deaths.

Error (fatality) data are available for the United States from the Bureau of Transportation Safety (U.S. DOT BTS), in conjunction with the U.S. Coast Guard. The numbers

*Table 2.1  U.S. Recreational Boating Fatalities and Analysis Using the DSM*

| | 1960 | 1970 | 1980 | 1990 | 1994 | 1995 | 1996 |
|---|---|---|---|---|---|---|---|
| Total fatalities (recreational) | 819 | 1418 | 1360 | 865 | 784 | 829 | 709 |
| Boats (1000's) | 2,500 | 7,400 | 8,600 | 11,000 | 11,400 | 11,700 | 11,900 |
| Mby | 2.5 | 7.4 | 8.6 | 11 | 11.4 | 11.7 | 11.9 |
| AccMBy | 2.50 | 9.90 | 18.50 | 29.50 | 40.90 | 52.60 | 64.50 |
| IR(F/Mby) | 327.60 | 191.62 | 158.14 | 78.64 | 68.77 | 70.85 | 59.58 |
| AccMBy | 2.50 | 9.90 | 18.50 | 29.50 | 40.90 | 52.60 | 64.50 |
| AR(F/accMBy) | 327.60 | 143.23 | 73.51 | 29.32 | 19.17 | 15.76 | 10.99 |
| AccMBy | 2.50 | 9.90 | 18.50 | 29.50 | 40.90 | 52.60 | 64.50 |
| CR(CR(0)=327) | 327.60 | 82.73 | 44.27 | 27.76 | 20.02 | 15.57 | 12.70 |

Data sources: U.S. DOT Bureau of Transportation Safety, 2000, and U.S. Coast Guard.

are presented in Table 2.1: there are now nearly 12 million recreational craft (boats), and each year several hundred people die from errors (accidents) in these technological machines. Again, we would expect human error to be a large contributor to the error rate, although we have not found a fraction stated.

Again using the DSM, we must take the accumulated experience as the basis. As with the large ships, we adopt the number of millions of boat-years (MBy), knowing that these boats do not spend all year in the water, nor are they used all the time. This analysis is then used to form the row in Table 2.1 for accumulated experience (accMBy). Only a fraction, f, of this total experience is the actual accumulated experience, or usage or risk-exposure time.

If we knew how long each boat was occupied on the water, we could use that value for f. But we do not know. If about 2–3 days were spent each year, that would be equivalent to multiplying the boat-years by a small usage or risk-exposure factor since the risk is higher for the shorter time than we have estimated. This fraction of the year actually at risk could be about 0.1–0.01 or 10 to 1% of the time.

The result of all the calculations and data shown in Table 2.1 is illustrated in Figure 2.12 where the IR, AR, and CR are plotted for the available data in Table 2.1. We can see that the AR is close to the CR, suggesting insufficient learning.

The exponential model fit to these data is simply given in Figure 2.13 by:

$$IR \text{ (per million boating years)} = 59 + 321 \exp(-MBy/12.7)$$

which implies a minimum death rate of order ~60 per million boating years.

Since the average fatality rate is about 60 per MBy, that is an error frequency of order $(60/(1,000,000 \times 365 \times 24)) \times f$, or about 1 every million boating-hours (accBh), assuming

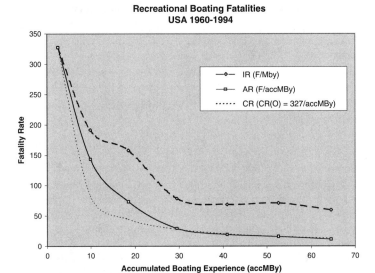

*Figure 2.12  Fatalities due to recreational boating in the United States since 1960. (Data sources: BTS, 2000; U.S. Coastguard.)*

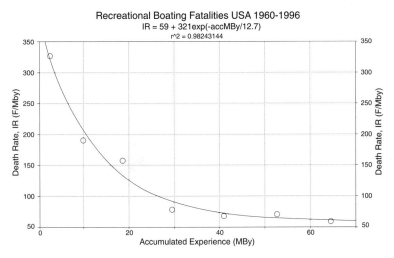

*Figure 2.13  The MERE exponential model for recreational boating in the United States.*

f is about 0.01 (~80 hours per year). This is a risk frequency of about one death for about every 150 years of actual afloat or boating experience, not knowing exactly the average number aboard each boat and for how long. There are too many assumptions to be sure, but the result is somewhat lower than the other numbers we have derived previously for other activities.

## The Effect of Limited Data: Smaller Countries and Less Experience

Again, we can show that the approach works on and for much smaller data sets, as we have shown before for aircraft near-misses. As an example, we have analyzed the fatal marine accidents for the United Kingdom from 1988 to 1998. The U.K. maximum experience over these 10 years is only about 5000 ship-years (5 kSy afloat), which is only 0.5% of the 1 MSy of world experience over the past 25 years.

The result is shown in Figure 2.14 for the fatal accident rate. The wiggles and the rise and falls in the IR and AR curves in Figure 2.14 are due to single accidents having a large effect, particularly on the IR. The error rate (losses) is now lower than that expected from the CR in 1988, and there is evidence of a learning curve. The IR fluctuates and is generally around the order of 10–20 per kSy, somewhat higher than the world data, but consistent with the lesser experience.

Thus, a mini-learning curve exists in the subsets of the larger world data, and we are averaging over larger and larger amounts of data as we group countries and regions together. That was also shown for the aircraft case, when we were comparing regional and world error (crash) rates.

So, *how safe are you*? All these data analysis cover more than 10 years of experience and some 70 million ship-years. The answer for ships and boats seems to be similar to

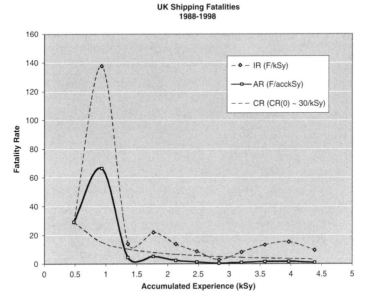

*Figure 2.14  The smaller experience data set from the United Kingdom for fatal accidents. (Data source: U.K. Department of Environment, Transport and the Regions, 2000.)*

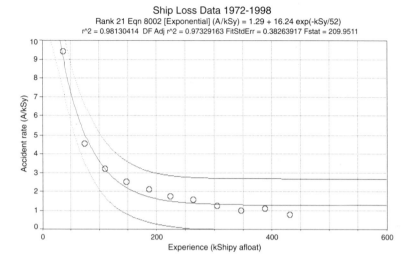

*Figure 2.15 Example of predicting the accumulated accident rate (AR) for world shipping out to 600 kSy using the exponential model.*

that for aircraft travel: a fatal or serious error occurs every few hundred thousand hours of risk exposure.

## Predicting Future Accidents and Loss Rates: Insuring the Risk

The data on the loss rates for the world shipping over 500 metric tons can be used to make predictions. In Figure 2.15, we show an example derived from the AR representation of the data (for the interval 1972–1998). The "best fit" exponential model was obtained by the commercial curve fitting software program TableCurve 2D.

The MERE line is shown in Figure 2.15, as given by

$$A = A_M + A_0 \exp(-N^*)$$

So, the loss rate for shipping is, as we by now expect,

$$\text{Loss rate, AR (A/acckSy)} = 1.3 + 16 \exp(-\text{acckSy}/52)$$

From this model, the best estimate of the lowest (minimum) rate predicted is ~1 per 1000 accumulated ship-years (acckSy), within the range of 1 in 20 times (i.e., at 95% confidence) of ~3 and zero. This is within the range of the IR value we derived before and is consistent with the value given by Pomeroy (of Lloyd's Register) at the Saarbruecken World Safety Congress.

This loss rate gives, with 35,000 vessels afloat, a minimum average attainable rate of ~35 per year. This number compares to our prediction based on a larger interval of ~75 per year, so we clearly have at least about a factor of 2 uncertainty in predicting future loss rates.

The future risk of losses is then known to within a factor, depending on the confidence level or uncertainty band chosen. Presumably, we expect that commercial insurance premiums in the rating process take into account this loss rate and the factor of 2 uncertainty estimate in the prediction.

Traditionally, of course, insurance estimates are based on the history and *prior year losses,* because current premiums are earned in the sense that each year must at least cover the losses (plus some profit!). There may be a variety of prior years that will vary according to the insurance company, and some companies may have their own data and also compare against competitive rates. That rating or premium setting approach inevitably means the premiums lag the learning curve, and also does not allow for accurate future predictions. This uncertainty in prediction must surely have something to do with the other accident that could not happen: the bankruptcy, even if only temporary, of many of the leading "names" (insurers) at the great insurance house of Lloyd's of London in the 1990s as large disasters caused unexpected losses.

The degree to which reinsurance (loss and risk spreading among multiple carriers) is used will also depend on the loss and risk ratios, and the cash flow due to premium inflows and loss outflows. Thus, Mutual Associations have been formed, which also conduct rigorous classification and safety and loss analyses, such as in the reports of the U.K. Property and Indemnity Club, and support extensive research on safety and the impact of technology changes. The DSM can also be used to make useful predictions and to determine the relative effect of changes in design and navigational practices.

### Technology Changes, But the Loss Rate Does Not

In his book *In Peril on the Sea*, which documents shipwrecks in Nova Scotia since 1873, Robert Parsons makes a key observation about the role of technology change:

> Shipwrecks . . . even considering modern advances in technology and communication are not a thing of the past. Humans, being inventive creatures, will devise and design aids to overcome problems; yet, despite every precaution that can be taken in designing navigational aids, human error and mechanical breakdown can still result in marine disasters.

This leads to a very important question we have all seen addressed in many ways in modern technologies: If the financial risk is unacceptable, or the accident or error rate is too high or does not change with time, what can be done? Can a change or significant advance in the technological system change the learning curve?

At first sight, this seems obvious, but it is not an easy question to answer. Changes often occur gradually, or data are not available for both the "control" groups of before (without) and after (with) the change. To disentangle reductions in errors when change or learning is still occurring and the rate is changing also complicates the analysis, to say nothing of the debate. We can see an analogy, for example, in the world debate over global warming and climate change: significant changes in climate are occurring naturally while the impact of human emissions due to technology shifts are superimposed or even interwoven.

We decided again to look for what the shipping data loss might indicate. Shipping has seen massive changes in technology over the past 200 years: from wooden hulls to steel vessels; from sail and wind power to steam and turbine power; from navigating by compass and maps to radar tracking and automated global positioning systems; from small merchant ships to supertankers. Did these shifts in technology change or reduce the shipwreck (error) rate?

We had the available data from Berman, who carefully recorded the shipwrecks in the United States from 1800 to 1971, some 13,000 in all by his count. What makes Berman's spectacular data record particularly interesting is that it covers this period of technology change, and his data include the number and date of the wrecks *and the age of the ship*. Thus, we can estimate the accumulated experience (in ship-years, Sy) for these very losses. To be consistent with our previous data analysis for more recent years, and to keep the task manageable, we totaled a sample of the losses for ships that were greater than 500 tons only. This gave a sample of some 500 losses, excluding those due to war and acts of aggression, with more than 10,000 years of accumulated experience to the point of loss.

The DSM uses a matrix of the form shown below for this case, where the arbitrary entries illustrated for each year are then totaled and analyzed to calculate the IR and AR.

| Year to Loss | 1801 | 1802 | 1803 | 1804 | 1805 | 1970 | 1971 | Total Losses | Ship Years | accSy |
|---|---|---|---|---|---|---|---|---|---|---|
| 1 | 1 | | | | | | | 1 | 1 | 1 |
| 2 | | 6 | | | | | | 6 | 12 | 13 |
| 3 | | | | 2 | 8 | | | 10 | 30 | 43 |
| 4 | | | | | | | | 0 | 0 | 43 |
| 5 | | | 1 | | | | | 1 | 5 | 48 |
| 6 | | | | | | | | 0 | 0 | 48 |
| 7 | | | | | | | | 0 | 0 | 48 |
| 8 | | 4 | | | 3 | | 1 | 8 | 64 | 112 |
| 9 | | | | | | | | 0 | 0 | 112 |
| 10 | | | | | | | | 0 | 0 | 112 |
| 11 | | | | | | | | 0 | 0 | 112 |

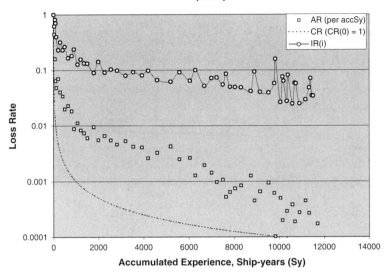

*Figure 2.16 Rate of shipwrecks over 500 tons in the United States since 1800 for 10,000 ship-years, derived from Berman's data and showing the IR, AR, and CR lines.*

The result of this DSM analysis of errors (losses) for a sample from this span of 170 years of calendar time with over 10,000 ship-years of accumulated experience is shown in Figure 2.16.

The best fit exponential learning curve is given by the standard MERE form:

$$IR \text{ (losses per Sy)} = 0.08 + 0.84 \exp(-accSy/213)$$

The loss rate falls quickly toward a minimum or asymptotic rate and is still slowly declining, although the minimum rate from the fit was not particularly sensitive to the sample size. There is no evidence at all in these data of a shift or change in the loss rate due to the major technology changes that occurred in ships from the 19th to the 20th century. That is what we know.

Presumably, the constant historical rate of shipwrecks is actually due to the factors that are not under anyone's real control: weather and human error, given the inherent hazard of being at sea and capable of sinking. Until a technology shift such as an effectively "unsinkable" boat is introduced and accepted, we may expect and predict shipping losses to continue.

It is worth noting that the asymptotic error IR (loss) rate of ~0.08 per Sy is then:

$$0.08 \text{ per } 365 \times 24 = 0.08 \text{ per } 8760 \text{ hours}$$

or 1 per 110,000 hours. This value is of the same order and extremely close to the minimum rate for near misses and aircraft accidents and is consistent with the range of independent estimates made for modern and recent shipping losses. *The minimum error rate has not changed in 200 years and does not shift from old sailing ships sinking to modern aircraft crashing.*

## 2.3 THE SAFETY OF ROADS: ACCIDENTS WITH CARS AND TRUCKS

Not all of us travel much, if at all, by plane or boat. But most of us in the industrial world travel by car, usually in a personal automobile as a driver or passenger. It is another example of a modern technological system with a control interface (or HMI), in which human control (the driver) interacts with a machine (the vehicle).

There are rules and procedures on how to drive and where, and signs and signals for us to follow. We drive to work, to the stores, to the beach, to the country, to visit relatives, and to vacation spots. As we drive we make errors while we are changing lanes, signaling, observing, passing, turning, stopping. We all have experienced having or avoiding an accident due to either our errors or those of other drivers. Police officials state that driver human error causes or is involved in about 85% of all automobile accidents.

Increased safety measures and devices, such as speed limits, seatbelts, and airbags, are all intended or designed to reduce the accident rate or risk (the chance of having an accident) and also the fatality rate or risk (the chance of being killed in any given accident).

But we know that as long as we drive around at high speeds in metal boxes on congested streets, collisions (accidents) will still occur. Errors (accidents) are so expected and commonplace that we are *required* to carry insurance against claims of damage and liability. The accident rate is measured by many competent authorities that specialize in statistics and safety reporting. Most countries have a government department dealing with transportation safety, and large amounts of data are available from the OECD, the U.S. Department of Transportation BTS, and the U.K. DETR. The data are followed closely by manufacturers and regulators alike to see the effects of road conditions, vehicle safety devices (seatbelts and airbags), speed limits, and driver alcohol level on accident rates.

*The United States has the most vehicles, the most experience, and the most fatal accidents.* In Figure 2.17 we plot the raw fatal accident rate statistics from the U.S. National Highway Traffic Safety Administration for 1975 to 1996, on a calendar-year basis. What are all these data telling us? How likely are we to have an accident or make an error? *How safe are you?*

The data in Figure 2.17 are for all 50 states and show a declining fatal accident rate with time, which recently has leveled off. Improvements in vehicle design (crashworthiness),

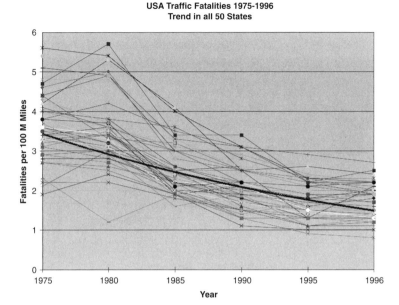

*Figure 2.17 The automobile death rate in all 50 states of the United States from 1975 to 1996. (Data source: Traffic Safety Facts, 1996 U.S. DOT NHTSA, Washington, D.C.)*

large-scale introduction of passive restraints such as seatbelts, and improved highways and freeways may all have contributed. But how much? Would a decline have happened anyway? Are some states "safer" than others?

Usually, the NHTSA and other such organizations report the data, analyze the declining rate, and use this to justify the continued introduction of safety measures, that is, legislation or standards aimed at reducing the fatal accident rate. There is about a factor of two variations in the rate between states, when shown in this manner.

To further examine potential state-by-state variations in experience, we were kindly supplied with 20 years of Australian data (by the Australian Transport Safety Bureau). The fatal injury rate was for seven states and territories, and we used the accumulated millions of vehicle-years, MVy, as the measure of experience since we did not know the distance traveled.

The data shown in Figure 2.18 follow a learning curve and have a MERE form given by

$$\text{IR (Deaths/1000)} = 0.11 + 0.32 \exp(-\text{accMVy}/1.35)$$

The minimum rate (~0.1) appears to have been reached in several of the states, and there is about a factor of 3 variation with increasing experience.

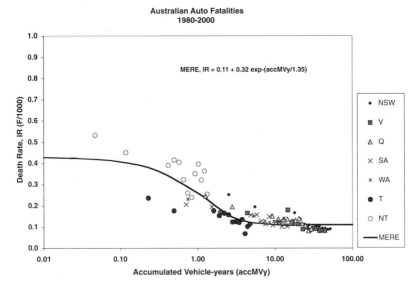

*Figure 2.18  The data for different states can be shown as a common trend, each point being for a year in a given state. (Data source: Australian Transport Safety Bureau, 2001.)*

## National Trends

In Figure 2.19, we transformed the data for the U.S. average (of all 50 states) into the fatal accident rates versus the accumulated experience, defined as the number of vehicles miles driven for the entire U.S. vehicle fleet. The units are hundreds of millions of vehicle miles (100 MVm) since the total time spent in a vehicle is assumed to be the time at risk as well as a measure of the possible extent of learning. For the 33-year interval shown (1966 to 1999), the number of vehicle miles accumulated in a year increased by almost 100% (twice the amount per year) as more and more vehicles were on the road, whereas the number of fatalities per year fell from nearly 50,000 to about 40,000. The traffic fatality data given in Table 1.1, in Chapter 1, showed how to calculate the numbers. Interested readers can now make their own plot using those data.

Clearly, the AR has fallen below the CR, there is evidence of a learning curve (cf. shipping where there was not), and the rate has fallen exponentially. But the fatal rate shows evidence of flattening out: the implication being that without a major shift in technology, for example to automated Intelligent Vehicle Highway Systems (IVHS) or remote vehicle control, there will be little further reduction. To check the trend, we were kindly supplied the data collected by the ATSB in Australia for 1980–2000. These data were also available state by state in slightly different form, and they also followed a learning curve.

From the U.S. data shown in Figure 2.20, we may deduce the MERE equation:

$$A = A_M + A_0 \exp(-N^*)$$
$$IR \text{ (F/100 Mm)} = 1.39 + 4.3 \exp(-\text{TrillionVm}/19.5)$$

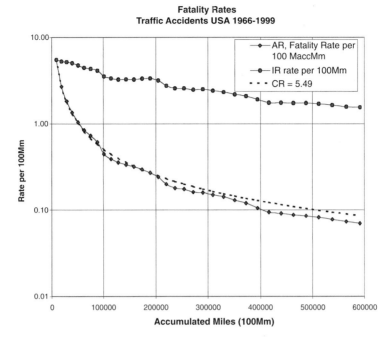

*Figure 2.19 The U.S. fatal accident rate and the accumulated experience for 1966 to 1999: the IR, AR, and CR in units per 100 million miles (100 Mm). (Data source: NHTSA, 2000.)*

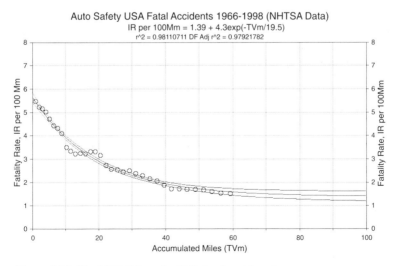

*Figure 2.20 The MERE fit to the U.S. fatal accident data.*

The projection shown is out to double the 1999 experience and predicts a minimum rate of ~1.4 per 100 million miles.

The average distance driven annually per vehicle in the United States was about 10,000 Vm/y from 1970 to 1996. The fatal accident rate both internationally over all the OECD countries and locally in the United States on a state-by-state basis is remarkably insensitive to the number of miles driven (i.e., the risk of a fatal accident is nearly constant for all the driving time). So a minimum fatal accident rate of ~2 per 100 MVm gives an average risk of an error (fatal accident) as about $2 \times 10,000/100,000,000$, or about 1 in 5000 years. For covering 10,000 miles per year driven at, say, an average of ~30 mph, we would drive perhaps 300 hours a year (or about 1 hour per day). So the frequency of a fatal error is

$$300 \text{ h/y} \times 5000\text{y} = 1 \text{ in every } 1,500,000 \text{ driving hours (1.5 Mdh)}$$

Now the measure of error is not just being killed; there are many more nonfatal accidents (what are called "dings" as well as collisions). The chance of a nonfatal accident in the United States is about 250 times more than that of a fatal accident (U.S. DOT BTS, 1999), illustrating the effectiveness of safety measures and the huge number of nonfatal accidents. So the error (accident) frequency is in fact 1 in 1,500,000/250, or 1 in 6000 driving hours. This frequency is at least 30 times larger than the airline error frequency. So what is the cause of this apparent insufficient learning?

The answer is the traffic density: on the one hand we are driving on very crowded roads, with random separation, while on the other we are flying in comparatively uncrowded skies where separation is controlled. The chance of an auto accident is largely dependent not only on the exposure in time in a given country (as we have just shown for the United States) as we drive around, but also on the density of the traffic in a given location. That is after all one of the reasons why insurance rates are higher in cities. If and when we make an error, the chance of a subsequent collision largely depends on whether other vehicles are around at the time.

This effect is shown clearly for the data for 1997 for 24 countries with 24 Bkm of travel in 1994, covering 522 MV, and nearly 100,000 deaths (see *Transport Statistics Great Britain 1997*, Tables 8.3 and 8.7, International Comparisons, HMSO, U.K.). The data in Figure 2.21 cover countries in Europe, Asia, and America, with accumulated experience varying by a factor of more than 100, between 28 and 3800 Bkm/y, totaling over 6 Bkm driven. Despite this large range, the fatal accident rate only varies within a factor of 3 with varying accumulated experience *and shows no systematic variation with country or accumulated experience.*

The graphs in Figure 2.22 also show this very clearly: the AR increases, and there is clearly insufficient learning with increasing accumulated experience!

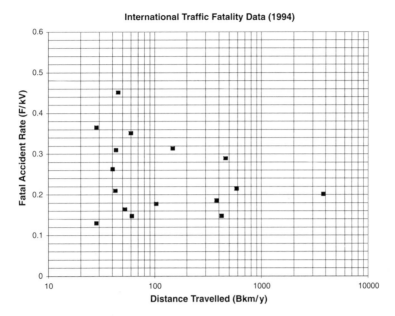

*Figure 2.21 The worldwide fatal accident rate as a function of traffic density and accumulated experience. (Data Source: **Transport Statistics Great Britain 1997**.)*

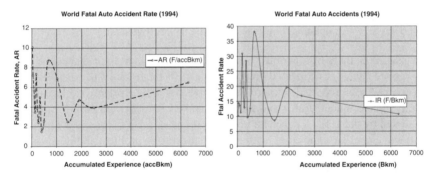

*Figure 2.22 The AR (on left) and the IR (on right) for world automobile deaths for 1994 as fatalities per Bkm (F/Bkm).*

Thus, we may conclude that, within a given country—as we have shown for the United States—learning with the technological system reduces the *fatal* accident rate. But the overall collision rate or average nonfatal accident rate is constant, mostly independent of learning and hence not affected by the accumulated experience. The average fatal accident rate throughout the world apparently depends only to first order on the traffic density and lies at a rate of about 0.2 per thousand vehicles in a given year (~0.2/kV/y) as shown in Figure 2.21. The accidents are unavoidable with today's vehicle and traffic technology; that could change with new automated traffic systems (ATS) that fix both separation and speed.

If you wish to avoid having an accident, of any type, then the best thing to do is to avoid driving in high-density traffic. *Thus, we say that drivers all over the world try not to have accidents (collisions) with about the same (non) success everywhere.*

What is more, the analysis shows that the error rate is now dependent almost entirely on factors over which the HMI has no control: the external factors or environmental factors of traffic density. Of course, the other factors of weather and road condition, visibility, number of lanes, average speed, etc., all now come into play. But the dominant factor is the error caused by the external density. *Increasing learning is no longer a major factor in these developed and industrialized, highly mobile nations.*

## Tire Failures in Transportation and the Concept of Critical Systems

We all use tires when traveling. Tires are not meant to fail without warning, but to run and wear to the end of their useful tread life, unless they are abused or punctured. But tires do and can fail, in many modes and for many reasons—like any technology, as a leading manufacturer explains:

> A tread separation is not always a tire defect. Rather, tread separations can result from a number of different causal factors related to tire use. All tires, no matter what brand of manufacture, can experience tread separations, and some will. This can happen whether as a result of low inflation, improper repair, a road hazard that causes a slow leak, an impact break or a similar incident. It is one of the most common failure modes for any steel belted radial tire, regardless of brand. ("Facts and Data Prove Wilderness AT Tires Safe"; Bridgestone-Firestone, Nashville, TN, July 19, 2001)

In the year 2000, there was an outcry over tire failures on sport utility vehicles (SUVs) in the United States. Unexpected accidents, some of them fatal and some leading to rollovers, had been reported for certain tire types made by Bridgestone-Firestone for multiple vehicle manufactures and factory installed on new vehicles. A huge recall was ordered by the U.S. Department of Transportation in 2000, for some 14.4 million tires. It was clear that some combination of manufacturing problem and/or stress under load in operation (driving), combined with the specific vehicle handling characteristics, had led to serious problems. However, it was not entirely clear where exactly "blame" should lie for the failures: the manufacturer for making them, or the vehicle maker for installing them, or some combination of the two. The two openly blamed each other in U.S. congressional testimony and statements, producing data and reports that showed either the tires or the vehicle as the more likely cause. The NHTSA started a large testing program on tread separation on all makes of tires.

Headlines are still running as a result of the significant impact on the industry:

> Firestone, other tire giants start safety push. (*USA Today*, April 6, 2001)

> Throughout our ongoing investigation of tires issues, we have shared all our findings with Firestone. We understand this has serious implications for Firestone, so we don't

take this action lightly, but we must act for our customers' safety. (Ford Motor Company Tire Replacement, May 2001)

... to find the whole truth regarding Ford Explorer rollover crashes, it is imperative that Congress, NHTSA, and the public examine the vehicle issues as well as tire issues. I have said from the outset that no research, analysis or remedy for tire-related Explorer rollover crashes can be complete without carefully addressing the contribution of vehicle characteristics. Today, I will present claim-data that show that the same tire on vehicles other than the Explorer performs quite well and that the tread separation rate, while still low, is elevated when that tire is on the Explorer. I will also present test data that precisely identifies that characteristic of the Explorer that makes it extraordinarily prone to rollover crashes in the event of a tread separation, an event that can happen with any tire. (John Lampe, Firestone Bridgestone Testimony, June 2001)

Answering the blame question alone could lead to a large product liability cost, over and above the cost of replacement, at $100 per tire, of some $1.5 billion if all were replaced. The recall statement reads in part:

The recall will cover all P235/75R15 Firestone ATX and ATX II tires (from 1991 to the present) and all P235/75R15 Wilderness AT tires (from 1996 to the present) manufactured at Firestone's Decatur, IL plant. Firestone does not plan to recall the approximately 5.6 million Wilderness AT tires manufactured at its other plants (Joliette, Canada and Wilson, NC) or other models of Wilderness tires. Firestone estimates that approximately 6.5 million of the tires covered by the recall (which include original equipment, replacement, and full-size, non-temporary spare tires) are still on the road.

The remedy will be to replace the defective tires with new Wilderness AT tires from the Joliette and Wilson plants, other Bridgestone/Firestone tires models, and competitors' tires. (Firestone, December 2000)

So, the implication is that some plants or processes were faulty and some were not, making the same nominal version of the tire. The error rate statistics can be calculated as follows. There were some 4300 reports of Firestone tire failure events listed in the U.S. DOT database as of January 2001. Assuming that all the errors have been found so far, and that not all the tires made have a problem since not all such tires have failed, this is an average error (failure) rate of

$$4300/14,400,000 = \text{one in 3350 tires, or } \sim 3 \times 10^{-4} \text{ per tire}$$

So far, as of December 2000, there have been 148 deaths reported in the DOT database. This fatality number depends on the type of failure and crash that occurs, as well as on the number of humans in the vehicle. The implied death rate is then ~3% of the number of reports, so that the overall death risk from the tire errors—whatever they may be due to—is

$$148/14,400,000 = 1 \text{ in } 100,000 \text{ tires}$$

Since the average auto is driven about 1 hour per day, on four tires, this is a death (error) rate of one in 25,000 hours for each and every day the vehicle is driven. That is close to the number for the fatality rate we have found for other industrial errors and implies that the risk to an individual is comparable between the two activities (working in a factory and driving on such tires).

For comparison, the total accidents attributed to tires in the United States were quoted by Goodyear Tire (www.goodyear.com), according to the NHTSA, as about 640 for 822 million tires in operation in the United States in 1999. This is a failure rate of 0.787 per million tires. Goodyear also points out very clearly that people should realize that modern tires are not indestructible: the *confluence of circumstances* causing failure can include overloading, underinflation, and general abuse, as well as the vehicle itself.

### Tire Failures in Other Technological Systems

To look at another technological system using tires, we reviewed what data we could find on the Concorde tire failure. A failed tire seemingly was an initiating event that brought down that great aircraft in the subsequent chain of events. The cause of the tire failure was claimed to be debris (a metal strip) on the runway penetrating the tire. Some 57 failures of tires had occurred before on the Concorde fleet; it takes off at over 200 mph. The causes were not all known. The analysis of these data in the BEA Concorde report and the events on takeoff only are in Table 2.2 as given in the report, where a failure is a burst or deflation.

We simply note that, using DSM thinking for these component failures, the rate falls as experience grows. Experience is measured in cycles accumulated, modifications also being made to tires and equipment. The apparent failure rate is also close to the failure rate for automobiles ($\sim 10^{-4}$), but we still need to know how much the tires were actually used. There are eight main-wheel tires per aircraft at risk during takeoff.

When we calculate event rate, the time in use (one cycle $\sim$2.7 hours flying, but only 30–45 seconds for actual high-speed takeoff), the average failure rate on takeoff is

$$(22/83,941) \times (3,600/45)$$
$$= 1 \text{ in every 500 hours of high-speed takeoff use}$$
$$\sim 1 \text{ in every 100,000 actual flying hours}$$

*Table 2.2  Tire Failure Data for the Concorde Aircraft*

|            | Takeoff Cycles | Events on Takeoff | Rate per Cycle |
|------------|----------------|-------------------|----------------|
| 1976–1981  | 24,052         | 13                | $5.4 \times 10^{-4}$ |
| 1982–1994  | 42,628         | 8                 | $1.9 \times 10^{-4}$ |
| 1995–2000  | 17,261         | 1                 | $0.6 \times 10^{-4}$ |
| Total      | 83,941         | 22                | $2.6 \times 10^{-4}$ |

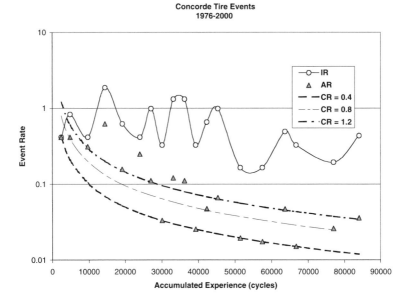

*Figure 2.23 The tire failure data per 1000 cycles for the Concorde aircraft. (Data source: BEA, 2002.)*

This last number is essentially identical to the failure rate for road tires as discussed above. At first sight, the first smaller number, which is based on the actual short take-off time, looks high. So we decided to compare the available tire failure data and risk, using the DSM and accounting for the very different tire populations and accumulated experience bases. That would help answer the question: are we safe in these different circumstances of driving on the roads or flying in a supersonic aircraft?

We see immediately in Figure 2.23 that, as often happens, the IR wanders around, exhibiting scatter and an unclear learning curve, having started from a rate of about 0.4 per 1000 cycles. When we compare to the AR, the picture is strikingly different. The rate "flip-flops" between the minimum rate limit of a low CR of about 0.4 per 1000 cycles, and a significantly higher rate of a CR > 1.2. The latest rate has increased again with increases in cycles: it is as if the failures fell to a minimum every now and then, and then rose. We do not know why, and this trend could be simply due to the difference between single and multiple tire failures, but clearly a well-defined learning curve was not being followed for this particular component.

The apparent minimum failure rate of 0.4 for 1000 cycles that was attained for the original tire type was a rate of about one failure for every 6750 flying hours.

Critical subsystem component failures are usually catered for by creating multiple failure paths with redundancies—often by using identical system components—thus allowing the overall target reliability for the system not to be compromised or allowing it to be achieved. Systems failure assessment is highly systematic and a well-documented process, but it is often judgmental, particularly failure rate assumptions and the assumed sequence and severity of events, relying on experience and lessons learned, for secondary failure effects. The overall failure effects can be minimized by systems segregation or by providing specific physical protection.

There can be critical single-point failures in a "system" that do not lend themselves to redundancy for practical reasons, for example, a tire blowout or jet engine turbine disc rupture. Design requirements do not allow such single-point failures to hazard the aircraft with catastrophic results. The likely severity of the failure and sequence can be assessed and sometimes tested. However, the analysis becomes increasingly difficult due to the possible number of interactive failure modes and effects. Tough design requirements, production, and maintenance standards have to be specified in order to ensure the overall safety targets are met.

In an ideal world, recorded system and component failure rates would be fed back into the original safety assessment analysis in order to check its validity and make corrections as appropriate. This requires accurate and complete reporting systems. Usually, corrections are made when there are obvious excessive failure rates involving critical components or systems, which will often include changes to operating procedures as well as mandatory modifications.

Failures and errors within these critical subsystems and components can then cause total failure of the system, of course. The identification of such so-called critical systems and components has led to a new licensing approach (as part of risk-based regulation). For example, for nuclear reactors licensed by the U.S. Nuclear Regulatory Commission, the design, verification, monitoring, management, and maintenance of critical systems have to be explicitly considered. However, this approach could become literally the search for the "needle in the haystack," as it is not possible to say which system will fail and when, and—as in the Concorde tire case—where the specific failure was already considered in the safety analysis. However, the actual *confluence of circumstances* had not been predicted.

### *Comparing Different Technologies*

We may summarize these apparently disparate data using the accumulated experience as a basis for the three very different applications: trucks, cars, and aircraft.

The data set represents completely different accumulated experiences (number of cycles/tires, C), which we would really like to compare on a common basis. Now, we do not actually know the number of tires used on Concorde to date, so we made a realistic estimate as follows, knowing the typical operating history. There are about 30 cycles for each tire life, and a total of eight main-wheel tires "at risk" for the entire total of ~83,000 cycles (TOFLs). That is:

$$83,000 \times 8 \sim 660,000 \text{ total tire cycles} = 660,000/30 \sim 22,000 \text{ tires used}$$

*Table 2.3 Tire Failure Rates in Autos and Aircraft*

| Application | Number of Events | Tires (M) | Failure Rate (F/M) |
|---|---|---|---|
| Concorde 1976–2000 | 57 | 0.022 | 2600 |
| Firestone recall 1991–2000 | 4300 | 14.4 | 299 |
| All United States 1999 | 647 | 822 | 0.787 |

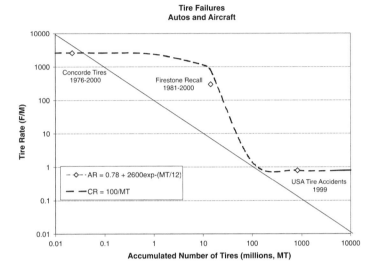

*Figure 2.24 Tire failure data plotted as the AR for different accumulated experiences and technological applications, and also showing the CR line for 100 failures per million tires. (Data sources: U.S. DOT, 2001; BEA, 2002; and Goodyear Tire Company, 2000.)*

The failure rate for 57 tire failures with 22,000 tires is then

$$57/22{,}000 \sim 1 \text{ in } 2600 \text{ tires}$$

So we took the actual tire failure rates shown in Table 2.3 for the Concorde, Firestone recall, and U.S. total, using the number of tires as the basis for the accumulated experience. We plotted the data according to a DSM analysis as the AR and CR, each per million cycles, as in Figure 2.24.

The CR shown is a typical value of 100 events per million tires, corresponding to a failure probability of $10^{-4}$. Using this AR plot, the MERE for these data is in fact given by

$$\text{AR (failures per million, F/M)} = 0.78 + 2600 \exp(-MT)/12$$

which result *illustrates a potential common learning curve for tire technology*. Tire technology is continuously improving as we learn, and new tire designs with more limited debris generation are now being adopted for Concorde aircraft.

## The Cost of Tire Recalls and the Legal Minimum Paradox

Tire recalls continue, and they offer an insight into the errors (defects) that may arise in modern technology and whether overall we are learning. Our DSM analysis also gives some insight into the relative error rates (in design, production, manufacturing, and operation) for a complex technological system where the human is involved in all of these stages. A major common problem in manufacturing is indeed determining whether a specific design, process, or plant is producing defective products or, as we would say, a manifestation of a *confluence of errors*.

According to the NHTSA and the courts, for the U.S. auto industry the legal definition of a "defect" (an error in our terminology) is that a vehicle or component is defective if it is subject to "a significant number of failures in normal operation." The determination of "defect" does not require any predicate of finding any identifying engineering, metallurgical, or manufacturing failures. Legally, "a significant number of failures" is a number that is "non de minimus," or more simply not the lowest that can be attained. But we know by now that to reach the minimum error rate (the "de minimus") we must trace our way down the Universal Learning Curve, starting out with higher error rates.

*So obviously, all manufacturers are logically and paradoxically at risk of initially having above minimum errors when the only way to achieve that lowest rate is to learn from having the errors themselves.* In such interesting ways does the legal system legislate, and we just call this a Catch-22 situation, the "Legal Minimum Paradox." The law can inadvertently discourage error disclosure, since errors when learning are always initially above "de minimus" and may be legally a "defect" in the product.

Tires fail all the time from different causes. The investigation by the U.S. NHTSA was on tire defects and tread separation for tires on large SUVs. The NHTSA concluded that tread or belt edge separation is the failure mode and that the chance of separation is more likely as the tire ages, towards the end of its life (3–4 years), and in hotter climates. The failure initiation and mode can be affected by details of the design and manufacturing process, specifically, the depth and size of edge pockets in the tread, reinforcing (wedge) belts, and even the curing and adhesion techniques.

The most recent study of defects in Firestone and other radial tires was included in an NHTSA report published in late 2001, which states:

> The focus of (the) investigation was on those non-recalled tires that are similar to the recalled tires; i.e., Wilderness AT tires of the size P235/75R15 and P255/70R16 manufactured by Firestone for supply to Ford Motor Company (Ford) as original equipment, as well as replacement tires manufactured to the same specifications ("focus tires").

> Most of the focus tires were manufactured at Firestone's Wilson, North Carolina (W2) and Joliette, Quebec (VN) plants, beginning in 1994. In late 1998, Firestone began producing P255/70R16 Wilderness AT tires at Decatur, Illinois (VD), and in mid-1999, it began

producing P235/75R15 Wilderness AT tires at a new plant in Aiken, South Carolina. Also, fewer than 100,000 P235/75R15 Wilderness AT tires were produced at Firestone's Oklahoma City, Oklahoma plant. The focus tires were predominantly used as original equipment on Ford Explorer SUVs and, to a lesser extent, on Ford Ranger compact pickup trucks, and as replacement tires for use on these and other SUVs and pickups.

. . . Therefore, on the basis of the information developed during the . . . investigation, NHTSA has made an initial decision that a safety-related defect exists in Firestone Wilderness AT P235/75R15 and P255/70R16 tires manufactured to the Ford specifications prior to May 1998 that are installed on SUVs. . . . The initial decision does not apply to the P255/70R16 tires . . . or any of the Wilderness AT tires manufactured after May 1998. (NHTSA Engineering Report, Office of Defects Investigation 2001)

These simple statements initiated the recall of a further 3.5 million tires in late 2001. They imply unacceptable specific error rates in design and manufacture at these factories and present a rare chance for us to DSM analyze the openly available manufacturing data given in the NHTSA report.

Manufacturers and insurers have been picking up the large costs of the claims due to the errors (tire failures, defects, recalls, and replacements). We all expect that tires should wear out their tread before they fail. If they do not, we may make a claim that there is a "defective product." What do these claim data tell us? Are different plants and tire types really different or defect-free? *Are you safe?*

## Tire Claim Rate Changes as Accumulated Production Increases

The latest data given in the NHTSA report are of two types: (a) the rate of driver/owner claims for tire defects as a function of the tire type, manufacture, and size; and (b) detailed examination of tire design and defects that arise in the tire structure itself. For Firestone tires, sufficient data were also given on the claims rate for three manufacturing plants (coded in Table 2.4 as VN with, VD, and W2) that had significant tire production output over the years of the recalled AT and ATX-type tires. These total production data in accumulated millions (accM) are shown in Table 2.4 for each plant, together with the claims rate per million tires (C/M).

The only assumption that we make is that the plants were learning to make both types of tire and that size was simply secondary. The claims rates, C/M, for all of

*Table 2.4  Production and Claims Rate Data for ATX and AT Tires*

| Plant and tire | VD ATX | W2 ATX | VN ATX | VD all AT | W2 all AT | VN all AT |
|---|---|---|---|---|---|---|
| Total (accM) | 2.45479 | 3.4981 | 4.471474 | 3.9546 | 5.2423 | 4.0157 |
| Claims (C/M) | 549.6 | 167.2 | 72.5 | 90.3 | 51.9 | 16.7 |

Source: NHTSA, 2001.

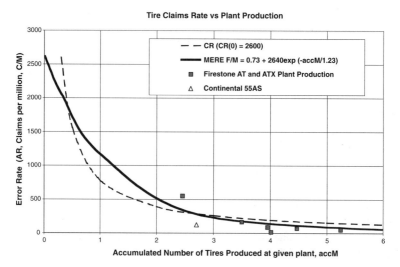

*Figure 2.25  Claims rate for three plants producing two tire types (squares) and for one additional manufacturer (triangle). (Data source: NHTSA, 2001.)*

the 15- and 16-inch AT tires are combined together and derived from the total production and claims rates, which were given separately for each size. The only other data point we found is from Continental Tires' response disputing the relevance of NHTSA data stating a claims rate of 124 per million for the 2.7 million Ameri550AS tires produced.

As always, we need a measure of the accumulated experience for the different production plants and tire types. Therefore, we choose the appropriate measure of accumulated experience as the total number produced of both ATX and AT tire types, in accM. We show the above error (claims) rate data as an AR plot in Figure 2.25.

The single triangle plotted is the lone data point given by the NHTSA in 2001 for the Continental 55AS tire for its accumulated production. We see immediately an apparent learning curve of decreasing errors (claims rate) with increasing production. In fact, the error (claims) rate is nearly a constant rate as the dotted CR line shows and is seemingly independent of the manufacturer.

The MERE line shown is from TableCurve fitting all the claims and failure rate data and is given by

$$AR = AR_M + AR_O \exp(-accM/K)$$

or

$$AR \text{ (Failure/Million, F/M)} = 0.78 + 2640 \exp(-accM/1.23)$$

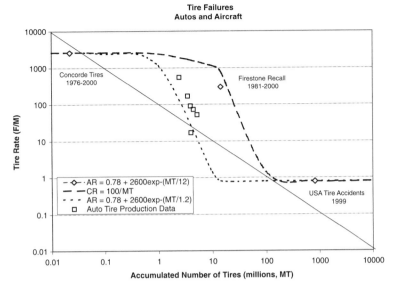

*Figure 2.26 Tire failure and claim rates presented using accumulated experience. (Data sources: U.S. DOT, 2001; BEA, 2002; and Goodyear Tire Company, 2000.)*

The plants are apparently on a steep and very nonlinear part of the learning curve and have not yet all reached the asymptotic minimum rate. It is implied that a plant production of about 1 million tires is needed to reduce the *apparent* claims rate—although statistically the plant may still actually have no claims in that initial interval, just as airlines may not have a fatal crash until accumulating ~6 million flights.

We replot all these claim rate data as squares and show the range of two MERE "fits" in Figure 2.26. This may be a controversial presentation, but it shows clearly that there may be a common learning curve. On the steep part of the curve, where the learning rate is highest, we do not have a great deal of data. The spread or space between the two lines can be taken as a measure of our "uncertainty," being about a factor of 10 for this tire example.

What is even more remarkable is that the implied initial error (claims) rate $AR_O$ obtained by back extrapolating the best fit exponential line to the claims data rate in Figure 2.25 could be comparable to the Concorde tire failure value of ~2600 per million! If true, then this is significant. We have already shown that the Concorde crash (started by a tire failure from an external cause) fits with world airline fatal-accident data. So it seems that the trend in the production error rate for the above tires mounted on SUVs is not inconsistent with the initial fatal accident rate on all the world's airlines for the past 30 years or so. The common factor in these diverse applications is clearly not the transport mode, production process, or service conditions, but is *the unavoidable involvement of humans in all the stages and uses of the design, manufacturing, production, and operation of technology.*

Thus we have shown, using the DSM, that:

1. The apparent rate of errors is tied to the accumulated experience (cf. air-lines).
2. The accumulated experience is suitably measured by the total production.
3. The extrapolated initial error rate could be comparable for tires in com-pletely different applications.
4. Apparent differences in claims rates between tires and factories can be a result of the differing accumulated manufacturing experience.
5. The differences can be analyzed using a DSM AR plot.
6. The apparent changes with time should but do not take into account the changing production quantities.
7. The "Legal Minimum Paradox" can apparently place all manufacturers at risk during the early stages of new production, and therefore inadvertently may not encourage error reporting and disclosure.

These are new observations implying the importance of factory-to-factory differences for a given design as we are learning from accumulated experience. If they are right, potentially there are some profound implications. Such analyses and results were not in the NHTSA report.

As always, it would be useful to have some more data points: we have analyzed and plotted what we found. There were insufficient data given on the quantity of tires pro-duced at different plants by other manufacturers for us to be able to analyze and assess the relative error (claims) rate for other tire types and designs. The data given by the NHTSA for other tires *with* claims are shown in Table 2.5, and we note that none are lower than the minimum given by the DSM of ~0.8 per million as derived from the U.S. total 1999 production.

We do not know or claim to have all the answers, and we really hope we are not in error. Nevertheless, the implication from the DSM analysis is that the apparent error rate assumed, used, or observed in setting production goals, insurance claims, and loss rates is highly dependent on the production experience and sample size.

**Table 2.5  *Claims Rate Data (C/M) for Other Tires on Specific Vehicles with Unstated Total Accumulated Tire Production***

| | | |
|---|---|---|
| Uniroyal Tiger Paw | P235/75R15 | 3.4 |
| Goodyear Wrangler RT/S | P265/75R16 | 2.3 |
| Goodyear Wrangler AP | P225/75R15 | 1.9 |
| Michelin XW4 | P235/70R15 | 1.9 |
| Goodyear Wrangler RT/S | P235/75R15 | 1.8 |
| Goodyear Wrangler RT/S | P265/70R16 | 1.3 |
| Goodyear Wrangler RT/S | P235/75R15 | 1.2 |

Source: NHTSA, 2001.

## 2.4  THE RAILWAYS: FROM THE ROCKET
##     TO RED LIGHTS AT DANGER

Railways are a great way to travel—they are seemingly very safe. Yet both tragedy and triumph mark the age of the railroads. The great engineers of the Industrial Revolution, George Stephenson and Isambard Kingdom Brunel, harnessed the power of steam and the strength of iron with their best ideas. The monuments to their craft and skill can still be seen in the replica of the "Rocket" in Manchester Museum, and in the great curved Tamar railway bridge in Devon. On the maiden run of the Rocket, the world's first scheduled train service, a bystander was killed. Some 180 years later, modern trains are still being derailed, running stop signs, and colliding.

### *Passing Through Red Lights (Signals at Danger)*

Trains travel on a section of track according to signals and switching of lines that are intended to ensure, for any single track, that there is adequate separation from other trains. If there is a possibility of a collision or inadequate separation, the train is switched to another section of unoccupied track, sent to a siding to wait, or shown red signal lights because it is dangerous to proceed. On seeing the red light, which is set on gantries by the track so it is visible, the driver should stop the train until it is clear and safe to proceed. Sometimes, the driver does not stop—either missing the signal or passing it, knowing it to be red.

This is a highly studied problem: in the lexicon of the railways, it is sometimes called a SPAD (a "signal passed at danger"). It has been found that train drivers with little experience, particularly those in their first year, are far more likely to pass a red light—commit a SPAD—than more experienced train drivers. The error rate declines with experience up to about 25 years. Then, as train drivers approach the end of their careers, there is a small increase. Disregarding the signal (human error) accounts for over 50% of the known cases, due to causes such as inattention, misreading, wrong expectation, or lapses of concentration. It has been pointed out by the U.K. Association of Locomotive Engineers and Firemen (ASLEF) that 85% of SPADs had been regarded as simply "driver error."

The crash outside Paddington Station in London, U.K., in October 1999 was an example of a modern accident, when a passenger train almost inexplicably passed a red danger signal before crashing into another train at high speed. There is an extensive report on the accident, "Train Accident at Ladbroke Grove Junction, 5 October 1999, Third HSE Interim Report," from which extracts are given in more detail in Appendix B.

> The accident occurred at 8:09 A.M. when a Thames Train 3-car turbo class 165 diesel train travelling from Paddington to Bedwyn, in Wiltshire collided with a Great Western High Speed Train (the "HST") travelling from Cheltenham Spa to Paddington. The accident took place 2 miles outside Paddington station, at Ladbroke Grove Junction. Thirty-one people died with a further 227 taken to hospital and nearly 300 people were treated for minor injuries on site. The initial cause of the accident was that the 165 passed a red

signal, SN109, and continued at speed for some 700 m before it collided with the HST. The closing speed was in the region of 145 mph. The reasons why the 165 (train) passed the red light are likely to be complex, and any action or omission on the part of the driver *was only one such factor in a failure involving many contributory factors*. (Emphasis added)

The phenomena of drivers passing through red signals, apparently unaware of their error is a known, but comparatively rare phenomenon. In the absence of an effective auto-mated system to prevent this type of failure, such as Automatic Train Protection (ATP), reliance upon drivers' correctly interpreting and responding to signals alone results in potential for a residual level of risk. (U.K. Health and Safety Executive, 2000)

The 300-page final report ("The Ladbroke Grove Rail Inquiry") contained 88 recommendations covering, as all such weighty studies tend to do, the causes and consequences of the SPAD, plus one covering the actual implementation of the recommendations. The crash was related to a *confluence of factors*:

- Problems with signal visibility, readability, and sighting in a multitrack complex
- Inexperience of a driver following inadequate standardized training
- Lack of continuity in management following prior events and recommendations
- Lack of integrated standing instructions (procedures) and guidelines
- Lack of effective automatic warning and communication systems
- Poor division of management responsibilities and accountability
- Potential conflicts among performance, cost, and safety

The report states variously in Sections 11.29, 9.67, 11.27, and Recommendation 33:

Error should, of course, be distinguished from a deliberate violation of a rule or proce-dure. It is essential that the industry redoubles its efforts to provide a system of direct management and training that is secure against ordinary human error whilst endeavour-ing to reduce the incidence of such human error to an absolute minimum. Mr. G.J. White, Inspector of Railways, HMRI, stated that in his experience "simply by recording driver error as a cause of the signal being passed at Danger is substituting one mystery for another one." The Group Standard on SPADs and its associated documentation should be reviewed to ensure that there is no presumption that driver error is the sole or principal cause, or that any part played by the infrastructure is only a contributory factor.

Thus, it is clear that the interaction between the human and the technological system causes the error, and that the whole system contributes in various unexpected and unforeseen ways.

It is also clear that train drivers pass through red lights just as car drivers do—some-times because they just can't see them, or are distracted, or miss them; are trying to go past the red signal too closely; and/or are in a hurry. What we see in these studies of the HMI and the errors is this same *confluence of factors*, some human, some mechan-ical, some design, plus other failures, that all conspire to cause the crash.

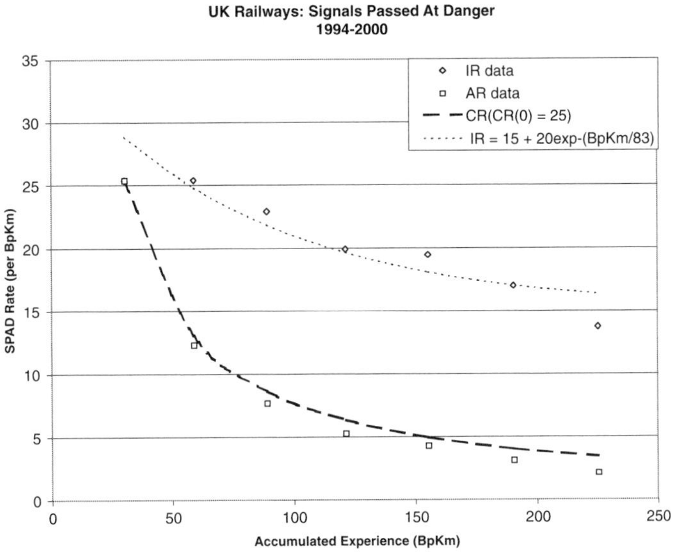

*Figure 2.27 The IR, AR for the U.K. data for trains inadvertently passing red signal lights, showing the MERE equation. (Data source: U.K. Health and Safety Executive, 2001.)*

One of the controversies behind this and other incidents was that the railways in the United Kingdom had been recently privatized. The immediate implication was that private enterprise had not done a good job. In fact Railtrack was placed in railway administration by the government, and the Railtrack Group no longer controls Railtrack PLC operational safety while in administration. What do the data say?

We examined the latest SPAD data history, which was available from the U.K. HSE Railways Inspectorate for the years 1994 to late 2000. There are many classifications of types of SPAD, but we assumed that just the act of passing a red light was, irrespective of the extent or consequences, the important error. For the accumulated experience, we took the passenger-kms (pKm) as a measure, although strictly we should have had the number of actual train-km of operation. The data are shown in Figure 2.27, where we extrapolated the data for 3 months to the close of 2000.

Although the data set is quite limited, we can observe that the IR does show evidence of a decrease due to learning and falls below the initial CR. The MERE fit for the exponential model is shown in Figure 2.27 as the dotted line and was derived by iteration to be

$$IR \text{ (SPADS per BpKm)} = 15 + 20 \exp(-BpKm/83)$$

This fit implies that the minimum rate is ~15 per billions of passenger kilometers (BpKm) as a prediction to at least twice the present experience with the present signaling and train technology.

This minimum rate can be converted to an error rate per hour if we know the average number of passengers per train and train speeds. Assuming for the purposes of an order-of-magnitude calculation about 100 passengers per train and an average speed of 30 km/h, we can convert the SPAD rate to the error rate per hour as follows:

15 per BpKm
> = 15 per 10 million train-km @ 100 passengers per train
> = 15 per 300,000 train-hours @ 30 km/h
> = 1 per 20,000 train-hours

This approximate rate is amazingly close to the rate for industrial errors, as we shall see later, and was not something we expected to find at first. A more precise estimate is obviously possible, but the inference is that the SPAD error rate is typically the same order as for the upper end of errors in today's technological systems.

### Train Derailments

Trains can come off their tracks, despite careful design and maintenance, and derailments are a major accident cause. There may be other consequences (hazards) too, as in an accident in Red Creek, Canada, in January 2001, when a derailment fractured a tank car full of ammonia. The fumes and potential hazard caused the evacuation of some 5000 residents in the area and made headline news for a day or two.

There is more government, media, and public attention these days, and this was certainly the case for the derailment accident in the United Kingdom where there were many casualties. We may quote from the inquiry "Train Derailment at Hatfield, 17 October 2000, First HSE Report," which states some apparently mundane but at the same time revealing facts:

On 17 October 2000, the 12:10 P.M. train (1E38) Kings Cross to Leeds passenger express train departed from Kings Cross, it was due to arrive at Leeds at 14:33 P.M. At 12:23 p.m., the train operated by Great North Eastern Railway (GNER), and, as far as we can ascertain at present, travelling at approximately the line speed of 115 mph, derailed roughly 0.5 miles south of Hatfield Station (approximately 16.8 miles from Kings Cross). No other trains were involved. Further details will become available once the signalling system's data tapes have been analysed.

As at 19 October, 4 passengers have been confirmed as dead, and 34 others suffered injuries. The derailment occurred approximately 16.8 miles (27 Kilometres) from Kings Cross station, on curved track between Welham Green and Hatfield in Hertfordshire. There are no points or signals nearby that are relevant to the derailment. Lines are electrified by the 25,000 volts overhead line electrification alternating current system. The maximum permitted line speed is 115 mph on the fast line and there were no speed restrictions in place. Early indications suggest that the immediate cause of the accident was a broken rail but this cannot be confirmed finally until all other avenues have been explored and eliminated. Findings to date are:

- there is obvious and significant evidence of a rail failure;
- there is evidence of significant metal fatigue damage to the rails in the vicinity of the derailment;
- the only evidence to date of wheel damage is consistent with the wheels hitting defective track;
- there is no evidence, so far, of a prior failure of rolling stock.

(U.K. HSE, 2001)

The result of this apparently normal accident was extensive public and political outcry, a major program of track inspection, and drastic speed restrictions on 300 sections of the railways. The resultant disruption of services and timetables drove many passengers onto the highways in their cars, a potentially more risky form of travel!

Such accidents make headlines and still occur regularly. On December 2, 2000, at about 5:00 A.M. a freight train was traveling on its own track in the Punjab, India, at the Sarai Banjara rail station when some four of the freight cars derailed. This "derailment" may occur when the turning iron wheels rise up over the edge of the track. This rising up over the edge can occur because of a combination of factors: excessive speed, bending forces causing sideways movement or inclination of the track because of insufficient track lubrication, or just too much vertical versus lateral force to keep the wheel recessed in the track. Sometimes this effect is due to loading, compounded by the makeup and order of the cars that alters the link forces between them.

The derailed freight cars spilled onto the adjacent track. Five minutes later, despite warning, and in the reduced visibility, a passenger train on its way to Ambala collided with this wreckage, smashing coaches, trapping more than 200 passengers and killing nearly 40 more.

As in the many accidents we have examined, multiple errors may be involved. *There is in fact an unforeseen confluence of contributory factors* that include:

- Exceeding safe speed limits
- Passing signals at danger (SPAD) or having no such signals
- Track failures
- Loading or wear, causing derailments
- Insufficient driver experience
- Inadequate stopping distance
- Poor communication and maintenance systems

Plainly, at this mature stage of railway technology, such accidents should not be happening. But they are, despite the safety measures, track inspections, training, signals, and other warning devices that exist. What do the accident data say? *Are you safe?* Are we learning?

Key data on accidents and injuries are available from the U.S. Federal Railway Administration, the U.S. and Canadian Departments of Transportation, and the U.K. Department of the Environment. We have examined these data. They cover both passengers and workers (employees), as well as those who expose themselves to the hazards of trains by trespassing unwittingly or wrongly on rail crossings or rail property.

An excellent review presentation has been published and kindly made available to us by the Association of American Railroads: "U.S. Railroad Safety Statistics and Trends" by Peter W. French, dated September 20, 1999. This is an exhaustive analysis of safety data of all types for the past 20 years of U.S. railroad experience. French clearly shows that the data support the facts, that not only have accident, fatality, and injury rates dramatically declined in that time (by 60% or more in many cases), but also there is a significant drop in employee injuries. The presentation includes these key remarks:

1. Railroads have dramatically improved safety over the last two decades.
2. Railroads compare favorably with other industries and transport modes.
3. The most troubling railroad safety problems arise from factors that are largely outside railroad control.
4. Railroads have implemented numerous and effective technological improvements and company-wide safety.

The "troubling" external factors referred to here in point 3 are mainly those accidents due to trespassing on railways, auto drivers not stopping at railroad grade crossings, and the influence of alcohol use. Obviously, the railroads cannot control the errors of others, as exemplified by these two types of external accidents.

On a comparative basis, the analysis by French also shows that railways on a per-passenger-distance-traveled basis are less hazardous than or comparable in hazard to other transport modes (Figure 2.28). These are all very encouraging trends.

The data are reported on a calendar-year basis, which does not fit the DSM approach or our earlier work, and the French analysis is also based on calendar time (years). So we have reexamined and recalculated the accident and injury rates on an accumulated experience basis, which we must choose on some defensible and, we hope, rational grounds.

## The Accumulated Experience Basis for Railways

*The accumulated experience basis we have taken to be the millions of railroad miles (MTm) accumulated by trains,* similar to that which we used for the road analysis. This is both for the sake of convenience, since it is reported, and because that corrects for yearly and other variations in railway usage, rolling stock, and rail traffic. The accumulated distance traveled is also a measure of the hazard exposure—if the trains are not moving, there should be less hazard or risk exposure.

Now railways and also ships carry a great deal of freight, so there is some more mileage, as well as more accidents and injuries, due to nonpassenger traffic. But the

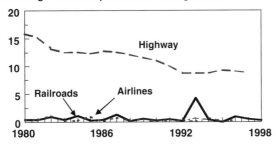

Passenger Fatalities per Billion Passenger Miles

Source: National Transportation Statistics, 1993, p. 52, 1996, p. 72
NTSB, Aviation Accident Data Base (special computer run)
FHWA, Highway Statistics, Table VM-1. NHTSA, Traffic Safety Facts, 1997, p. 18.
FRA Accident/Incident Bulletins, Tables 13, 36
FRA, RR safety Statistics Annual Report 1998, Tables 1-1, 1-3.
FRA website: http://safetydata.fra.dot.gov/officeofsafety/

*Figure 2.28 Comparison of passenger fatalities on a rate per distance-traveled basis (F/BpKm). (Source: French, Association of American Railroads, 1999.)*

experience basis with trains is being accumulated on an entire system basis—not just on passenger trains. So we might expect the learning rate, if it exists, to depend on the total accumulated experience, whereas passengers should have injuries only on passenger trains. The numbers are not too dissimilar.

Similarly, employee injuries (which far exceed those for passengers) and train derailment events should also depend on the total accumulated experience on the tracks. So we examine the total error rates (accidents, injuries, and derailments) based on the entire accumulated experience base.

Following the lead of the Federal Railroad Administration (FRA), we have taken the comprehensive data files for the United States, which were available to download for the years from 1975 to 1999. Over this 25-year interval, the United States has accumulated some 16 billion train miles (26 billion km), some 87,000 derailments, 27,000 fatalities, and over 800,000 injuries. These are quite small numbers compared to the equivalent road data.

### Accident and Error Rates

We calculated the fatal accident rates for all fatalities for the total experience base. The rate measure "passenger train mile, pTm" is based on the number of passengers, p, times the number of train miles, Tm (Figure 2.29). This measure tries to account for the effective hazard for all passenger travel and is consistent with the approach of the Railroad Association. However, we also analyzed the fatal accident rate for "total train miles," since this is the physical distance actually traveled by trains and is the exposure hazard for the actual rolling stock.

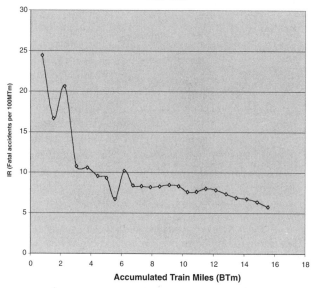

*Figure 2.29 The IR rate for "passenger train miles" (per 100 MpTm) for the United States railroads from 1975 to 1999. (Data source: U.S. Federal Railroad Administration, 2000.)*

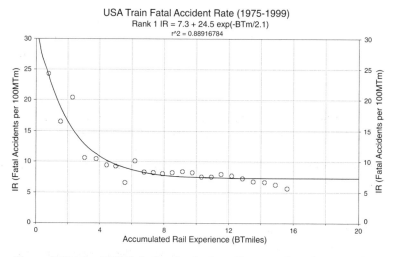

*Figure 2.30 The MERE fit for the fatal accident rate based on passenger train miles.*

The MERE for these data, which was selected by TableCurve, is shown in Figure 2.30 and is given by

$$A = A_M + A_0 \exp(-N^*)$$

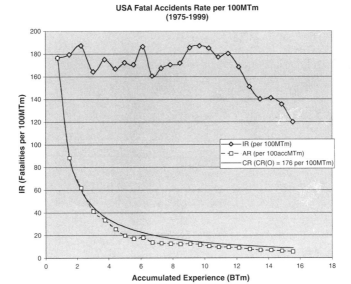

*Figure 2.31 The IR, AR, and CR fatal accident rates per 100 MTm for U.S. railroads based on total accumulated experience. (Data source: U.S. FRA, 2000.)*

or

$$\text{IR (Fatal accidents per 100 MpTm)} = 7.3 + 24.5 \exp(-\text{BTm}/2.1)$$

Thus there is evidence of a learning curve, and a decline to a finite minimum fatal rate that is of order ~7 per 100 MpTm. But we know this, to some extent, is an artifact of the "experience" assumed. In fact, we are examining all fatalities, not just passengers, where the number killed would depend on the train and on the passenger load (as it could for shipping, for example).

Reanalyzing the same fatal accident data but on the basis of the total U.S. railroad system train miles, we have a different trend (Figure 2.31). Basically the rate is changing slowly, with a decrease evident below the constant initial rate CR(0) from 1975. But the decrease is some 60%, not the factor of 5 to 6 evident in Figures 2.28 and 2.29 on a passenger-mile rate basis, simply because about 1000 people, both staff (employees) and passengers, are being killed each year on the U.S. railways.

## *Derailments*

Derailments, the other major accident category, present a significant hazard and potential economic loss. Depending on the cargo being carried, there are the supplementary hazards of explosion, chemical release, and/or fire. Obviously, the chance of injury depends on whether and how many people are on or are struck by the derailed train (as in the recent

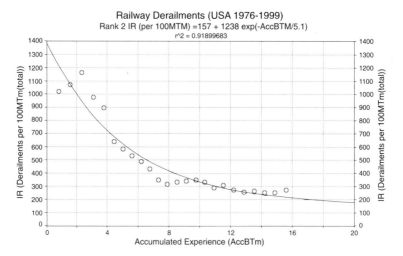

**Figure 2.32 The IR for derailments in the United States.**

India accident). High-speed derailments are a cause of great concern in several countries and are usually the result of human faults in track design, maintenance, and inspection, sometimes combined with errors in train speed, train loading, or axle and wheel condition. With freight and equipment loss, the cost of one derailment can be as high as many millions of dollars.

The U.S. FRA data for derailments are shown in Figure 2.32 for the total train miles (per 100 MTm). There is clear evidence (exponential decay) of a learning curve. The analysis conducted by Peter W. French, Association of American Railroads, showed a 70% reduction from 1980 to 1998: the DSM analysis in Figure 2.38 shows nearly a factor of 5 (80%) reduction since 1975 on an accumulated-experience basis.

The MERE best fit from TableCurve is also shown in Figure 2.38 and is given by

$$A = A_M + A_0 \exp(-N^*)$$

or

$$\text{IR per 100 MTm} = 157 + 1238 \exp(-\text{accBTm}/5.1)$$

so that the asymptotic minimum rate is estimated by the DSM as ~157 per 100 MTm and is *predicted* to occur at >20 BTm, which is quite soon, perhaps 5 years or so in the future.

Interestingly, the average fatal accident rate from Figure 2.31 lies between 180 and 120 per 100 MTm, with a minimum of 120/100 MTm. We can use these derived and observed error rates (fatal injury and derailment) to estimate the minimum error frequency.

We know how many miles (Tm) are traveled: we also need to know or estimate an average train speed. The error (fatal accident frequency, F) is given by the formula

$$F = (\text{IR per } 100 \text{ MTm}) \times (\text{Average train Speed, S, in mph})/100,000,000 \text{ hours}$$
$$= \text{IR} \times S/10^8$$
$$= 120 \times S/10^8$$

Now we need an average speed, so we assume about 50 mph for a train (we could take any other reasonable value). The error frequency is

$$F \text{ (total)} = 120 \times 50/10^8$$
$$\sim 1 \text{ in } 20,000 \text{ hours}$$

This estimate is about a factor of 10 higher than the minimum we have estimated before (an error interval ~200,000 hours) but lower than that for the roads (which was about 6000 hours).

### Other Limited Data Trends and the Debate on Risk Perception

The data for the European railroads (the EU) are also available (Figure 2.33). They also show an exponential decline in the fatality rate when expressed as passenger fatalities per billions of km (Bkm). Unfortunately, these data were given on a calendar-year basis again from 1970 to 1996, so without choice we assumed a roughly constant experience rate per year and plotted them anyway.

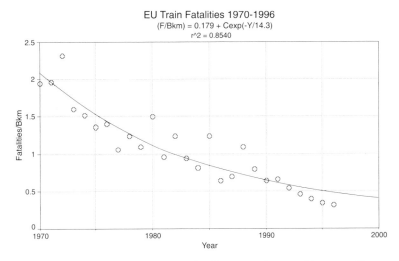

*Figure 2.33 EU data for passenger fatality rates. (Source: European Transport Safety Council, June 1999, Brussels, "Exposure Data for Travel Risk Assessment: Current Practice and Future Needs in the EU.")*

The MERE curve gives a minimum rate prediction of ~0.18 F/Bkm, which is comparable to the value of ~0.3 F/Bkm in the U.S. data reported by French. The fatal error frequency from this estimate on a passenger risk basis is then

$$F \text{ (passenger death)} = \text{Average speed in km per hour}$$
$$S \times IR \text{ per Bkm}/1{,}000{,}000{,}000 \text{ hours}$$
$$F = S \times IR/10^9$$
$$F = (80 \times 0.2)/10^9$$
$$\sim 1 \text{ in } 60 \text{ million hours (Mh)}$$

This frequency is very much less than the exposure from the overall error frequency from other sources of error and is clearly negligible. The chance and risk of any individual (that is, you or me) being killed on a train is small, comparable to that of being in an aircraft accident, as shown in Figure 2.28.

What do all these numbers tell us? What we have apparently shown using the DSM is that the fatal error (death) risk is very small for rail passengers, which is what the industry reports and states; but overall the major error frequency (which include death of employees and/or derailments) is high. *Thus, the apparent risk perception could be high (accidents and injuries occur quite often) while actually the individual risk is extremely small.*

Is this a key distinction? Perhaps the overall risk of an accident or error is a measure of the societal perception of risk, whereas in quite contradictory fashion, the individual passenger risk is very low. This potential for confusion or misperception would at least help to explain recent U.K. attitudes toward rail safety, and the public and political clamor to reduce accident rates and to greatly increase safety measures.

A similar argument holds for the Concorde and the aircraft data that we have analyzed. Any individual's risk of being killed in a Concorde plane crash is very small, and the error (crash) frequency for Concorde, we have shown, is the same as that for all other aircraft, when based on the accumulated experience. *Nevertheless, the political and public perception of risk was such that, because of one major crash, certificate of airworthiness withdrawal and safety improvements were required, which meant the literal temporary grounding of this technology.*

# 3

# WORKING IN SAFETY

*". . . whoever deprives himself of pleasures, to live with anguish and worries, does not know the tricks of the world, or by what ills and by what strange happenings all mortals are almost overwhelmed."*

—Niccolo Machiavelli

## 3.1 INDUSTRIAL ERRORS AND CHERNOBYL

According to the United Nations International Labor Organization (ILO), more than 330,000 workers are killed each year in accidents at work. People expect to be safe in their work, not to be exposed to risk and danger, unless they have a dangerous profession. We looked at the data for both the safest and the riskiest industries. We have carefully examined the almost completely safe industries with a premium on safety to see what the lowest achievable rate might be. We have chosen exactly the ones discussed by Perrow (1984): the nuclear and chemical industries, because of the potentially serious consequences of accidents, where extra attention should be placed on safety, for protecting workers, process equipment, and the public. We used the DSM and compared the findings for these very safe industries with "normal" industries to see what might be the difference in error rate. Since no industry is error-free, as we are now proving, it is important to determine what the premium on safety is and what actual change is made in the error rate.

### Errors at Three Mile Island and Chernobyl

In 1979, operators at the Three Mile Island nuclear plant in the United States ignored and misinterpreted instruments and warnings that indirectly indicated the reactor was losing water from a valve that was stuck open. Believing the reactor to be still full, they turned off the emergency cooling system. As a result, the partially dried out core overheated, the nuclear core partly melted, and the nuclear industry in the United States was set back some 20 years.

The many and various inquiries all showed the now all-too-familiar lists:

(a) Inadequate training
(b) Violation of procedures
(c) Control room information deficiencies
(d) Malfunctioning of equipment
(e) Generally, lack of management attention and safety culture

The strong concrete containment stopped much radioactivity release, but the power company was discredited and no order for a similar plant design (made by a reputable boilermaker and naval supplier) has occurred since.

The "lessons learned" were extensive, and major training, back-fit, safety, and management programs were put into place at great cost. Plainly this accident should not have happened, and certainly it should not happen again.

Some 7 years later, on April 26, 1986, outside the town of Pripyat, in the middle of the night, the operators of the Chernobyl nuclear reactor prepared the plant for a test. For reasons they did not completely understand, it was difficult to control the reactor at the lower power needed, so many control rods were withdrawn and some of the emergency safety systems overridden. This Russian-designed plant was run like a chemical plant: operators used the extensive instruments and procedures to guide their actions. But there were no procedures in place for this test and no training.

At 1:23 in the morning, to start the requested test, the flow to the steam system was decreased, and half the reactor pumps were turned off (tripped). The shift chief, Alexander Akimov, was watching the instruments. He saw the reactor power increase unexpectedly, so he pressed the "scram" button to shut it down. It operated as designed for normal conditions, briefly inserting power rather than removing it. Within a minute, the reactor exploded, blowing apart the reactor vault, lifting the reactor head, and destroying the surrounding confinement building. The operators knew something had happened. Unable to believe the now-wild instrument readings and the sounds of the explosion, they sent someone down to actually look. He returned to the control room a few minutes later, his face red from radiation, and confirmed that something terrible had indeed happened.

Now uncontained, the reactor burned for 10 days, spewing radioactivity in a plume carried by the wind over large regions of the Ukraine and Byelarus, and leaving traces around the globe.

Heroic efforts put out the fire. Major work was done to build a new shelter structure around the smoldering hulk. Now, even 15 years later, the repercussions are still felt.

Some say the accident and its aftermath contributed to the fall of the Soviet Union, partly because of the fiscal and political strain of the accident just when glasnost (openness) had become the new political policy. Since that time, no more reactors of that design have been built. The Ukraine has promised to close the remaining Chernobyl plant many times, and an international effort to build a large structure to contain the ruins and debris is now underway. Although there is dispute over the number of deaths that might eventually be attributed to the accident, some 30 workers are known to have been killed.

As public fear of contaminated foodstuffs caused panic, Italy put its reactor program on indefinite hold, leading to its policy of eternal dependence on importing its energy needs (gas and nuclear electricity) through pipes and wires from the rest of Europe, Asia, and Africa.

The causes of the Chernobyl accident are many; the *confluence of circumstances* again becoming clear in hindsight:

- (a) Lack of training
- (b) Inadequate procedures
- (c) Violation of safety rules
- (d) Design weaknesses
- (e) Management failure
- (f) Pressure to complete

All this happened, along with the immense repercussions, in the United States and Russia—nations that could put astronauts into space, launch nuclear warheads, and challenge the world and each other for technological and political leadership.

Finally, in December 2000, many years later, came this report:

> Another malfunction forced Chernobyl-3 to be shut-down on 6 December, nine days before the plant is scheduled to permanently close. A safety system detected fumes in the reactor's forced circulation department. Workers struggled to repair a steam generator in order to bring Chernobyl-3 back on line. The repairs were expected to be completed on 12 December, just three days before the plant's closure.

The reactor will be reconnected to the grid on December 13, with full capacity being attained on December 14. A ceremony, with government officials and foreign visitors, is planned to mark the closure of the plant on December 15, 2000 (Associated Press, December 8, 2000, press release).

The plant was then closed with much fanfare and at great cost. Was this accident unusual or just another manifestation of error? Was the event typical of the technological systems we have designed and use every day?

### *Industrial Accidents*

We would all like to work in safe conditions, but accidents can and do happen. Not all of them are headliners; simple accidents and events occur every day. Legislation in force in the industrialized nations includes many measures meant to ensure safety, or at least to reduce risk. These include enforcing or using many of the following elements in one way or another:

(a) Standards on the safety of the workplace
(b) Operator training requirements
(c) Analysis and treatment of hazards
(d) Labeling of facilities and equipment
(e) Submission of safety and environmental reports and analyses
(f) Use of safety equipment and safe working practices
(g) Establishment of inspection and licensing regimes
(h) Use and wearing of personal protective equipment
(i) Restrictions on working hours and conditions to reduce fatigue
(j) Reporting and tracking of accidents and injuries in the workplace
(k) Tracking of hazardous substances in manufacture, storage, transport, and disposal
(l) Adherence to all rules, regulations, guidelines, and standards that are applicable
(m) Use of physical safety measures to avoid accidents (barriers, relief valves, etc.)
(n) Management training and leadership skills
(o) Rewards and incentives for safety
(p) Comparative best practices and benchmarking
(q) Dedicated safety staff and supervision
(r) Periodic reviews, refreshers, and updates
(s) Trade associations and industry safety testing
(t) Labor and union standards for worker qualification
(u) Insurance and independent outside reviews
(v) Safety audits, inspections, and walk-downs
(w) Safety newsletters, literature, signs, and brochures
(x) Maintenance standards and manuals
(y) Safety procedures and practices
(z) Plant and equipment design features

. . . and many more such approaches, techniques, initiatives, tools, and methodologies, with penalties for noncompliance with the laws.

*Industry generally believes that a safe workplace not only is good for the employees and those who work there, but also is good business.* A safe workplace is more productive, there are fewer lost hours due to injuries and outages, and the plant tends to operate smoothly; there is pride in the workplace, and maintenance supports safe operation.

In addition, the training and support provided often prevents unnecessary accidents. *All these measures promote a "safety culture," which emanates throughout executive management, line management, and staff attitudes and echoes in improved worker performance and expectations.* We shall return to some of these topics later, but even if we do all this, key questions remain:

(a) What is our error rate?
(b) How safe are we in the workplace?
(c) Can we expect accidents and injuries not to happen?
(d) What can be done to improve?
(e) Can we measure safety performance?

Many government agencies gather and study so-called industrial accident data, and reporting of this data is quite extensive. Employers and trade associations also track such data, for their own plant or industry. If the industry is perceived as a "high-risk" one, then more attention is often paid to it. Some industries have their own regulator, so that specialists can review and license the details of the equipment and plant design and operations. We authors are specialists from two such industries, aircraft and nuclear, where the regulators have a large say in what is done, how it is done, and by whom.

*But it is the operators and owners of the plant and equipment—the licensees and only the licensees—who bear the full and heavy responsibility for the safe operation of their facilities.* In a sense, the regulators insure the public against undue risk, requiring evidence of safe design and operation before allowing the plant and/or equipment to operate.

Table 3.1 lists the 31 industries with the highest rate of cases of nonfatal industrial injuries and illnesses in the United States as given by the U.S. Department of Labor.

**Table 3.1  The 31 Industries in the United States with the Highest Rates of Cases of Nonfatal Industrial Injuries and Illnesses (numbers are employment in thousands)**

| | |
|---|---|
| Metal doors, sash, and trim 32.4 | Hardwood dimension and flooring mills 36.9 |
| Meat packing plants 149.4 | Transportation by air 1196 |
| Gray and ductile iron foundries 25.1 | Fabricated structural metal products 467 |
| Motor vehicles and car bodies 343.7 | Food and kindred products 1686 |
| Truck trailers 39.5 | Furniture and fixtures 532 |
| Ship building and repairing 105.9 | Household appliances 117 |
| Iron and steel foundries 131.6 | Manufacturing 18 |
| Mobile homes 76.1 | Nonresidential building construction 635 |
| Automotive stampings 114.8 | Industrial machinery and equipment 2207 |
| Steel foundries 29.1 | Construction 5949 |
| Metal sanitary ware 14.9 | Non-durable goods 7598 |
| Commercial laundry equipment 5.7 | Agriculture, forestry, and fishing 1815 |
| Fabricated structural metal 80.9 | Plumbing fixture fittings and trim 24 |
| Poultry slaughtering and processing 248.3 | Mining 588 |
| Transportation equipment 26.61 | Vitreous plumbing fixtures 9.9 |
| Construction machinery 93.6 | |

Source: http://www.insure.com/business, "Industries with the most injuries and illnesses, 1998."

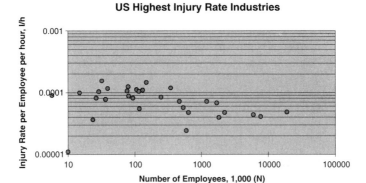

*Figure 3.1 Data for the injury rate for "high rate" industries in the United States in 1998. (Data sources: OSHA BLS, 1999; insure.com, 2000.)*

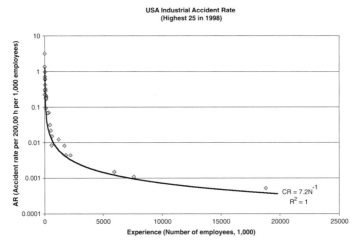

*Figure 3.2 The AR and the CR for the 31 highest injury and illness rate industries in the United States in 1998. (Note: on a logarithmic plot for both axes, the CR is a straight line of slope −1.) (Data source: U.S. DOL, 2001.)*

As we can see from Figure 3.1, there is not much variation between these "high rate" industries. In fact there is very nearly a constant injury rate (CR(I)) for all industries. We can now show two key points by using the DSM approach. In Figure 3.2 we plot what is effectively the AR, defined by the U.S. Department of Labor as the incident rate per 200,000 hours worked per 1000 employees, versus the size of the industry in terms of the numbers employed (in thousands) as given in Table 3.1.

The data do indeed follow the CR, where in this case it has the value of CR(0) ~ 7 per 200,000 hours per 1000 employees. For these data and these industries:

(a) There is no difference between the rates that cannot be ascribed to the differing sizes of the industries (i.e., the larger ones have only an apparent lower rate)

(b) The actual rates are the same, and there is no learning or benefit from any increase in the size of the industry as measured by the number of employees

*In other words, you and I are as safe working for a small industry as for a big one*. This is true at least for the higher risk examples discussed and classified as such by this sample from the U.S. Department of Labor. Industry size is not, then, an indicator of individual risk.

The error frequency (total injury rate) per hour worked in these industries from Figure 3.1 is on average ~1 in 20,000 hours, with a minimum value of perhaps 50% more. This average value is at least in the range of our other risk interval numbers, being exactly the same as that derived above for the total fatal injury rate for the railroads. We did not know of this similarity between these disparate and different activities beforehand: it was discovered by using this DSM analysis, exactly as we have described it.

*Is this coincidence—or not?*

## Underground Coal Mining: Accident Data Sources and Historical Analysis

Excellent historical data sets exist for fatal accidents in underground coal mining. We consulted and used the information available from the following open sources:

1. United States—U.S. Department of Labor, Mine Safety and Health Administration 1931–1998*
2. India—Journal of Mines Metals and Fuels 1901–2000
3. Australia—Queensland Department of Mines, Mines Inspectorate 1902–1998*
4. Australia—NSW Journal of Mines Metals and Fuels 1990–2000
5. South Africa—Chamber of Mines 1951–1999*
6. United Kingdom—Pit Disasters 1853–1968
7. Poland—State Mining Authority 1945–1999*

The data files marked with asterisks (*) are downloadable to various degrees over the Internet. The data sets together cover immense evolution and changes in mining technology, with major spans in time frames, production, employees, cultures, exploration, working practices, management, and national regulation and enforcement. They also represent a chance to see how the results and trends may be applicable industry-wide, and how the economic and competitive pressure to produce more may affect safety. We made spreadsheets of the data (in Excel (.xls) format) to analyze the trend.

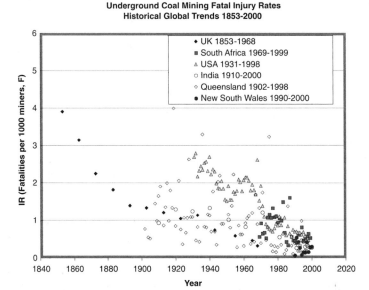

*Figure 3.3 The trends in coal mining fatalities over the past 200 years show a large decrease and remarkable similarities in death rates throughout the world.*

We can make at least three observations from the two centuries of data shown in Figure 3.3:

*Observation 1:* The rate has declined from ~4 per thousand miners in the early 19th century to a minimum today of between 0.1 and 1 per thousand at the start of the 21st century. All the data sets that show clear learning curves—a declining rate with time—and the clearest curves are for those with the longest time frames (e.g., for the United Kingdom and Poland). Shorter or smaller data sets with fewer miners (e.g., in Australia) show more statistical scatter, but still have the same well-defined declining trend.

*Observation 2:* The other remarkable feature is that the different data sets have similar rates and now are all close together at the lower values. Thus, across national borders the technology, mining systems, and safety management practices must at least be similar, if not identical. *Coal miners underground everywhere try to avoid accidents with about the same degree of (non) success everywhere.* Where you are exactly in accident rates depends on where you are on the learning curve; and the rate of decrease of accidents is dependent on the past history (the trajectory) and on the present practices (the rate of decline).

*Observation 3:* The rate of decline in the accident rate with time (year) approximates an exponential curve, initially falling away quickly, and then slowing to a steadier and slower decline, before almost leveling off, indicating that a minimum rate may be being reached.

Underground mining by its nature might be expected to be dangerous, but as the table of 31 industries shows, it is no different from other high-error industries. The data for average fatal injuries in the United States for all industries is given on a yearly basis from 1969 to 1997 (a time span of experience of nearly 30 years) by the U.S. DOL OSHA.

The average minimum rate is lower than the highest rates deduced above, and hence is about 1 in 50,000 hours (i.e., 2 per 100,000 hours). This value, which is within a factor of 4 of the value we estimated for the airline industry, implies that some industries may be nearer to a minimum error rate than others.

To use the DSM, we need a measure of the accumulated experience. The U.S. coal mining industry has an outstanding data and accident record, covering 1931–1998, available from the U.S. Department of Labor Mine Safety and Health Administration (the MSHA). We have analyzed these data using the DSM, adopting the millions of worker-years (accMWy) as the accumulated experience basis. We can show conclusively that the minimum accident rates are attained with a well-defined learning curve. Fatalities are of order less than 1 per million hours worked; and injury (error) rates are ~7 per million tons of coal production or ~56 per million hours worked. We have found similar excellent data, covering 1969–2000, available from the Chamber of Mines of South Africa for both gold and coal mines, and from Queensland's Department of Mines and Energy Safety and Health Inspectorate for 1882–1999. When analyzed with the DSM, all these data also show clear but not identical learning curves as shown in Figure 3.4.

Basically, the 20th-century mining data show similar but not identical transnational error rates and learning trends. This is evidence of some commonality of approach to

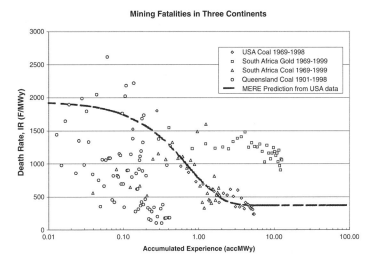

*Figure 3.4 Twentieth-century underground mining error data compared to MERE prediction for the United States.*

errors in this same technology (or industry) across the three continents. The MERE line shown for comparison in Figure 3.4 is the exponential model for the U.S. data for 1978–1998 and is given by

$$\text{IR (Fatalities per MWy)} = 372 + 1570 \exp(-\text{accMWy}/0.74)$$

Human error is a major contribution in complex mining technology.

## 3.2 ALMOST COMPLETELY SAFE SYSTEMS

Both the chemical and nuclear industries, like the aircraft industry, pay great attention to reducing errors and accidents. This is partly good business practice: a safe and well-run plant produces more because it spends less time out of service for maintenance, repairs, or just the pure hassle of the consequences of events. Attention to safety is also key to keeping injuries down, as well as the potential for injuries. Therefore, having an acceptable risk and an excellent safety record is as important as, and may well take precedence over, excellence in performance and production in order to achieve and maintain public, licensee, and regulator acceptance.

Today industry strives to achieve all these fine goals; however, it is clear that economic pressures rate high in the survival race of a privatized and globally competitive world. This has led to the perception and criticism that profits are being obtained at the expense of safety. No company is deliberately trying to get a poor reputation, real or imaginary, and put itself out of business; therefore safety is taken very seriously. Some major accidents have led to increased emphasis on top management's accountability for corporate safety standards and the introduction of safety management procedures, however they are defined.

Amalberti (2001) in his extensive writings and studies calls these technological activities the "almost completely safe industries" (ACSI). In these technological systems, errors (accidents) are reduced to small chances, where regulations and controls are highly enforced and adopted, and where the consequences of a large accident are really unacceptable to everyone. Amalberti also argues that these ACSI also require a different and more sensitive approach to safety management and regulation. Accidents in ACSI have been reduced by design and operation to very low levels of both chance and consequence. Therefore, operators and management should not be distracted by extensive safety measures or reporting to the regulator over operational minutiae, or take safety for granted. This is a fine balancing act and demands a safety culture of attention to detail without losing sight of the bigger picture and the overall "Safety Management" system.

The safety regulator, too, acting on behalf of society, must also not overprescribe safety requirements. We have shown that people tend to regulate what they know can happen so that, too simplistically, it "cannot happen again." But occurrences and serious events are a *confluence of events* in unique sequence, each item of which we may have already studied, but without anticipating the exact sequence that actually occurred.

Hence, *both designers and regulators adopt the "defense in depth" principle*. If one barrier or safety system is breached or broken, then there is another to protect the system against further problems, and another, and another, layer on layer. In this way, for example, relief valves to avoid excessive and dangerous overpressure in vessels are backed up by safety valves. To avoid "common" modes of failure, these safety and relief valves may be of different types by different manufacturers and may be powered or actuated by multiple, diverse, and independent systems.

What chain of errors can happen in design, or safety, or operation? As an example, at the Bhopal chemical plant in India, the inadvertent opening of a relief valve in a pressurized vessel enabled toxic chemicals to kill and injure hundreds of surrounding people. Again, at Three Mile Island, a relief valve had stuck open and showed that it leaked, but was ignored by the operators because of false indications and misinterpretations. The consequences, as for the other incidents, were extensive litigation and reimbursement, and an extensive public and fiscal liability for the operators. So clearly, operating safely is in everyone's interest. But are we or can we be completely free of errors?

If we relax, pay less attention, learn less, or forget, then we may find an increasing error rate. This is a precursor of things to come, since some of the errors may together form an unexpected confluence at some time, with very serious repercussions. That is precisely what happened at the JCO plant in Japan.

## 3.3 Using a Bucket: Errors in Mixing at the JCO Plant

It was Friday morning, September 30, 1999, and three workers were nearing the end of their shift. They wanted to complete mixing the batch of chemicals that day, so they would not only complete the work on schedule but also train others on the job starting the following week. They had already changed the mixing process, as agreed by the company safety committee, and changed the procedures.

Usually more than 100 operations had to be done, cross-mixing the liquids between many small bottles to ensure that all eventually contained about the same mix of liquid. Now they could use a big tank to mix them together and use a bucket to fill the tank. That was much faster and easier than using lots of smaller bottles.

The liquid was a solution of uranium for making nuclear fuel. The JCO plant and the process were regulated and licensed in Japan. The plant and procedures were all designed to avoid having too much liquid in one bottle. More than a certain amount of uranium solution in one container could achieve a self-sustaining uncontrolled nuclear reaction (a critical mass). This was dangerous, a well-known hazard, and calculated and understood—except that the workers had not been trained on this aspect. After all, the plant operation and procedures were all designed so that *having a critical mass could not happen.*

Unknown to the workers and the regulator, the new mixing procedures with a larger vessel violated the rules and license, since it could allow a larger batch to be mixed more quickly. So this was done regularly. Now a new batch of solution was to mix with a different (higher) amount of uranium. That day, to further speed up the process, the workers took the solution in the bucket and tipped it into an even larger process vessel, as that would speed up the mixing even more. One worker asked the supervisor if it was safe to do this: since it had been done rather like this before, the answer was both affirmative and wrong. The mass in the vessel exceeded the critical limit, a self-sustaining nuclear reaction occurred, and two workers were fatally injured by the radiation, dying later in the hospital. They really had no idea that their actions would not only endanger themselves, but also others. The relevant quote is as follows:

> In the JCO accident, information related to the violation of operational rules plays an important role in the derivation of background factors affecting their wrong decision-making. The wrong decision to be analysed is the use of the precipitation tank instead of storage column. The following reasons may make this decision rational:
>
> (1) Since: (a) the pouring of 16 kg-Uranium into the storage column was usually done, and (b) the precipitation tank and the storage column had the same capacity, the use of precipitation tank seemed to have no problem.
> (2) The work with the precipitation tank is easier to handle and not time-consuming.
> (3) The insignificance of using precipitation tank was confirmed by the supervisor, who seems to have mistaken the job of high-enriched uranium for the ordinary low-enriched one. The reason for his mistake may be because: (a) the last order of the product was issued three years ago, or (b) the treatment of high-enriched uranium was unusual.
> (4) Three workers wanted to finish the job in the conversion building quickly because: a) the crew leader wanted to keep his promise to give the process co-ordinator a sample of product just after the noon break on September 30, (b) they wanted to start the training of newcomers at the beginning of the waste process-ing procedure, or (c) the working space was small and located far away from the main building.
>
> Among the above reasons, the first three ones explained why the precipitation tank was used instead, while the last one shows the purpose of three workers' actions. From the only viewpoint of fulfilling their purpose to finish the job quickly, the selected action is reasonable because it can reduce the working time. The problem is that they seem not to have paid any attention to the side effects of their actions on safety, or a criticality accident at all in the evaluation of their alternatives. Generally speaking, people do not take an action leading to bad results if they know them. The benefit of the short cut in their operation cannot be comparable to the loss caused by a severe criticality accident. As the crew leader testified, the danger of their work or the criticality accident seemed to be out of consideration. (Kohda *et al.*, 2000, PSAM 5, Osaka, Japan)

As news of the accident spread, panic was in the air, and confusion occurred between the local and national governments. People in the surrounding area were advised to stay indoors in case excessive radioactivity had been released. Neighborhood evacuations were ordered. For 24 hours, staff at the JCO plant worked to stop the reaction in the vessel as quickly and safely as possible. False rumors spread that the roof had been blown off. By the next day the sheltering requests for people to stay indoors were canceled at scattered times. Emergency management was literally made up as the events developed, since there was no coherent emergency plans. In the aftermath of the incident, new regulations were invoked, the plant was closed, the management indicted and convicted of causing death, and the public scared.

The subsequent extensive official inquiries in Japan, when completed, were exhaustive and publicly self-shaming for all those involved. In our terminology, the investigations showed that again we have a *confluence of errors*:

(a) Failure to follow safety rules
(b) Absence of training
(c) Inadequate oversight and management attention to safety
(d) Violation of safe operating procedures
(e) Time pressure to complete ("pressonitis")
(f) Lax regulation, inspection and enforcement
(g) Inadequate process monitoring
(h) Poor process and safety system design
(i) Focus on production and costs

These have been generically described as failures and errors under the heading of a lack of the needed "safety culture" or a "learning environment," which is our preferred expression.

What a learning environment means in practice is the existence of an ideal total work environment that strives to be safety conscious in every aspect. The whole work-related system emphasizes and pays unending attention to safety, in all aspects of design, operation, management, and rewards. Thus, the management, organizational structure, staff training, plant condition, trust, free communication, open reporting, blameless appraisal and self-criticism, awareness and readiness, and pay raises all constitute a "culture" that reinforces and rewards safe operation by all.

So what do the data tell us? Are these ACSI really safe? How much safer are they than other industries? How can we tell if they are paying sufficient or insufficient attention to safety? In fact, *how safe are you*?

## 3.4 CHEMICAL AND NUCLEAR PLANTS

The chemical industry pays close attention to safety, and several of the major companies are seen as leaders in industrial safety. After all, they deal with explosives, highly toxic and inflammable substances, poisons and pesticides, acids and alkalis, all in large quantities. Many of these "chemicals" are potentially or actually lethal to humans and/or the environment. In large plants in the old Soviet Union, for example, in the north Russian town of Nickel, the use of heavy metals has destroyed the local area for many miles and the average life expectancy is low. The discharge or inadequate storage and disposal of chemicals have sometimes caused health and property damage, as at the infamous Love Canal in New York. There are Superfunds to help pay for the cleanup of the careless legacies of the past.

We all use many chemicals in our daily lives, without harm and generally without damage. They are an integral part of our modern society. The industrial production of chemicals is now highly regulated in many countries. In the United States, the Environmental Protection Agency (EPA) was formed basically for this reason: to measure, monitor, and regulate discharges and their effects. Lists of dangerous substances have been developed, and the required training is enforced by EPA and by the Occupational Safety and Health Administration (OSHA). Violations of limits and standards are punishable by fines and restrictions on operation. With this Damoclean sword situation, no wonder the industry is involved in safety and the reduction of errors.

We were kindly supplied with some data from the United Kingdom by the Chemical Industry Association (CIA), which carefully tracks the industrial injury rate. The data set was for 1970 to 1997, covering some 6 million worker-hours. The instantaneous error (accident) rate, IR, is plotted in Figure 3.5, against the accumulated experience in hours worked. Since we did not have the exact numbers, we have approximated the hours worked each year from numbers given for the estimated workforce.

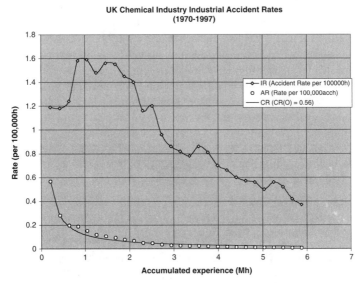

*Figure 3.5 U.K. chemical industry accident rates. (Data source: U.K. Chemical Industries Association, 1998.)*

It is clear that a learning curve is present, with an almost steady decline from year to year with an increase, looking like an occasional slight bump or lump in the rate, about every 1–2 Mh or 5 years or so. They look similar, almost as if each bump creates a new learning curve so that every time the rate goes up, the industry brings it down again over about the next 2 years. Overall, the rate has dropped by an impressive factor of 3 or 4 (some 300 to 400% reduction) over the nearly 30 years of accumulated operation.

The MERE line or Universal Learning Curve for a selected data subset is given by TableCurve as the exponential equation

$$A = A_M + A_0 \exp(-N^*)$$

or

$$IR \text{ (per 100,000 h)} = 0.1 + 10 \exp(-accMh/0.7)$$

Thus, this selected data set shows possible evidence of having an initially high rate that falls away with experience. The asymptotic or minimum error rate predicted in the far future by the DSM is then ~0.1 per 100,000 h (plus or minus ~1 at 95% confidence). The minimum rate observed *today* is a frequency, F, of one every 200,000 hours, plus or minus ~0.4 at 95% confidence, exactly the same value we obtained for fatal aircraft crashes and for near misses.

*Thus, these two completely and totally independent but "almost completely safe industries" (ACSI) of aircraft and chemicals have the same minimum error rate today.*

This rate is some factor of 4 to 10 lower than that for the 31 high-risk (error rate) industries and apparently demonstrates the effectiveness of safety management systems and techniques. It also shows that the minimum rate is not zero, at least based on what we can see from 30 years of data, more than 200 years of history, and many millions of hours and flights operating, using, and developing technological systems.

## Nuclear Industry Events

Another industry with a premium on safety is nuclear power, the generation of electricity using the splitting of the atom. This process causes so-called fission products to form in the reactor, which are toxic and highly radioactive. So they must be contained within physical barriers, by the fuel, by the casing for the fuel, by the high-pressure circuit, and by the concrete and steel containment building. The focus is on defense in depth, to guard against errors and accidents with these multiple barriers to minimize any activity release and possible damage.

Since the consequences of a large release could be very serious, the plants are licensed by regulators and operate with very strict conditions and limits on reactor power and the physical state of the plant. Many diverse, redundant, and independent safety systems are also deployed, and standards for design, manufacture, maintenance, and operation are at the highest level.

The regulators enforce these standards and track management and plant performance, as well as setting safety limits. The internationally respected U.S. Nuclear Regulatory Commission (the NRC), for example, follows and grades the performance of operating plants and publishes these on the Web, using performance indicators based on the availability and smooth operation of the key plant safety systems. Several of the indicators, but not all, demonstrate or are following learning curves when the accumulated experience is expressed as the integrated power output from operation.

The chance of a serious accident is very small (in technical jargon called extremely remote), but two have occurred (Three Mile Island and Chernobyl) in designs that were different from those most widely deployed. They both displayed decision-making errors, nonadherence to procedures, and a *confluence of factors* that included significant errors in design and operation.

Thankfully, such large accidents are indeed rare events, but smaller incidents that are called "significant events," just like industrial accidents and injuries, occur more often. These events may be reportable to the regulator, but in any case—just as in aviation—tracking the events is seen as an important activity.

The events range from the trivial to the serious. So databases of "significant event reports" (SERs) are collected and maintained, analyzed, and examined. The industry really wishes to reduce these events to a minimum, for safety and performance reasons, and tracks and records the events, classifying them according to the system involved and the root cause. The tradition of the U.S. industry operations is also largely derived from the U.S. nuclear navy, with a strong emphasis on accountability, rigorous training, and adherence to procedures.

In France the expert adviser to the nuclear safety authority, the Institut de Protection et de Sûreté Nucléaire (Nuclear Protection and Safety Institute—IPSN) examines operating events, just as in the commercial airline industry Mandatory Occurrence Reporting and Flight Operations Quality Assurance (FOQA) systems. IPSN has requested the French plant operator (Electricity de France, EDF) to systematically report SERs, known there as "significant incidents for safety." This has provided a rich database that IPSN analyzes, since EDF operates 58 nuclear reactors and has on average ~400 SERs per year.

IPSN, as reported by Baumont *et al.* (2000), has also developed various methods for analysis of operational feedback for lessons learned and databases for incident analysis. As described by Baumont *et al.*, the databases focus on technical or human factors and on the description of the causes and the consequences for facility safety to reduce the frequency, causes, or consequences. Since quick and correct recovery also improves safety, IPSN developed the "RECUPERARE" method with the objectives of:

- Identifying mechanisms and parameters that characterize events
- Classifying deficiencies and associated recoveries
- Classifying events according to previous parameters
- Making trend analyses

A metal strip lying on the runway...

...Punctured the tire on the Concorde
so it disintegrated on takeoff...

...Causing this aircraft to crash nearby

*Simple errors can lead to large consequences, as in the sequence of events for the crash of the Concorde Flight F-BTSC, which started with the puncturing of a tire by a strip of metal debris left on the runway. (Photos courtesy of: Bureau d'Enquetes Accidents, 2001, and Ralph Kunadt, 2001, www.airlinerphotos.com.)*

When a sudden loss of oil lubrication...

...Caused a loss of main engine power...

...The cargo ship drifted and collided
with a shop and hotel complex

*Sometimes a sequence can be simple with unexpected results: loss of engine power
due to failure of a simple engineering system. In this case, failure of oil lubrication
led to the ship drifting into the hotel on shore. (Photos courtesy of: U.S. DOT NTSB
from report PB98-916401 NTSB/MAR-98/01 1996.)*

A typical serious accident risk is
when one vehicle...

...For whatever reason collides with others

*Collisions between trucks and automobiles occur on regulated highways and are due to a confluence of human errors in dealing with routine changes in road conditions and the actions taken at traffic signals and in turning. (Photos courtesy of U.S. DOT NTSB and NHTSA.)*

A radial tire is made with steel belts...

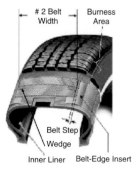

...Which sometimes can form a small defect

...And lead to the entire tread separating

*Small defects that may arise in manufactured components can lead to failure, as in this tire case, and to massive recalls of the product at great expense. (Photos courtesy of U.S. DOT NHTSA from Engineering Analysis Report and Initial Decision Regarding EA00-023: Firestone Wilderness AT Tires, U.S. Department of Transportation, National Highway Traffic Safety Administration, Safety Assurance, Office of Defects Investigation, October 2001.)*

Although the signal was at red...

...The driver did not see or obey it...

...Resulting in a collision and fire

*Human error is also involved in missing signals such as red lights: in this case the railway signals suffered from poor siting and visibility at a complex set of junctions, leading to a head-on collision. (Photos by permission of U.K. Metropolitan Police from the Ladbroke Grove Rail Inquiry, HSE, 2001.)*

Unqualified use of a key from this box...

To change these points on this track...

...Caused two trains to collide...

*Sometimes an apparently innocent act, like changing the points on the track to let a train through, can lead to unexpected consequences, in this case a head-on collision. (Photos courtesy of Australian Transport Safety Bureau, ATSB, from Rail Investigation report "Collision Between Freight Train 9784 and Ballast Train 9795," Ararat, Victoria, November 26, 1999.)*

Chernobyl reactor operators at this control panel...

...Misinterpreted the instruments—now covered in plastic sheets

...And destroyed the reactor, as seen from a helicopter

*Errors in conducting tests on an operating plant, without adequate training or procedures, can lead to significant accidents, as in the case of Chernobyl Unit 3. (Photos courtesy of Romney Duffey, private archive.)*

When the broken mangle in the laundry
was repaired with an ungrounded welder...

...Fire spread to where ropes were stored

...damaging a modern Fantasy Liner

*Simple maintenance errors at work can lead to large problems: in this case lint ignited
from the arcing of an unearthed welder during repairs in the laundry, causing fire to
spread through ducts and the ship. (Photos courtesy of U.S. DOT NTSB from report
PB2001-916401, NTSB/MAR-01/01, 1998.)*

When the black waste fuel in this tank...

...Was too thick and plugged both
the relief and safety valves...

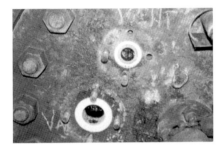

...Heating overpressurized and blew it up

*Pressure vessels can fail, and humans at work try to make things happen, in this case trying to pump and heat a viscous and partly solidified but flammable liquid, which had already blocked the safety relief valves on the tank, causing it to explode. (Photos courtesy of U.S. DOT NTSB from report PB2001-917002 NTSB/HZM-01/01 DCA99MZ004, 1999.)*

Reportedly, fire caused smoke to...

...Spread to the cockpit of a jet airliner...

...Which ultimately crashed. The wreckage has been reconstructed as a part of the thorough investigation into the causes of the accident

*Fire and smoke are known flight hazards: not all materials used are fireproof under all conditions, so a fire can start out of sight and spread, as reported on a jet airliner that ultimately crashed. (Photos courtesy of Transport Safety Board (TSB), Canada.)*

A landslide caused the dam to overflow

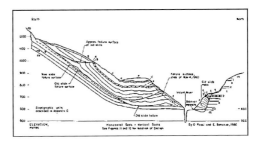

That caused a village downstream...

To have over 2600 people killed...

*Errors in dam siting and construction can occur, leading to failure: in this case the dam overflowed because of an unexpected slide-induced wave. (Photos courtesy of U.S. Dept of Interior from Bureau of Reclamation Dam Safety Office Report DSO-98-05, 1998.)*

When unexpected heavy rains fell...

...The rail track was undermined...

...Leading to an expensive derailment

*When undetected by inspection, errors in construction and monitoring can lead to failures, as in this case when unexpected high rains caused undetected flows that undermined a railroad bridge. (Photos courtesy of U.S. DOT NTSB from report PB98-916303 NTSB/RAR-98/03, 1998.)*

When a new house was built, electric
and gas lines were laid too close...

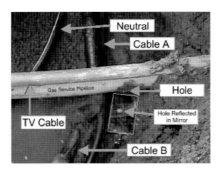

...And when the gas pipe was holed...

...The explosion blew up the houses

*Laying gas and electric pipes and then uncovering them again is common practice,
but errors leading to pipeline damage often can lead to gas explosions, in this case
destroying homes. (Photos courtesy of U.S. DOT NTSB from report PB2001-916501
NTSB/PAR-01/01, 2001.)*

When a low loader truck...

...Became stuck on this rail crossing...

...A train hit it and was derailed

*Unless railroad track and roadway level crossings are eliminated, errors causing collisions will continue to happen, as in this case of a truck becoming stuck on the crossing in the path of an oncoming train. (Photos courtesy of U.S. DOT NTSB from report PB96-916201 NTSB/HAR-96/01, 1996.)*

When the nut jammed on the screw thread...

...Some of the control surfaces jammed...

...And the plane crashed in the sea

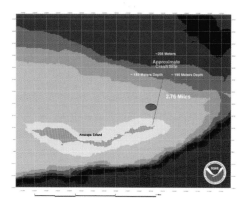 

*Ideally, to avoid errors, there should be no possible "single points of failure" in a system design that can lead to serious events, like this crashing of an aircraft. (Photos courtesy of U.S. DOT NTSB, 2001.)*

Following a rocket launch...

...The vehicle became unstable...

...Because the adapted fins separated,
either due to flight loads...

...or from damage from rocks
in sandbags used to protect
the base of the launcher

*Apparently small design or operational changes can cause accidents, in this case where the modified tail fins on a rocket failed during launch, causing it to crash. (Photos courtesy of Australian Transport Safety Bureau, Final Report of the Investigation into the anomaly of the HyShot Rocket at Woomera, South Australia on 30 October 2001. Australian Transport Safety Bureau June 2002, BO/200105636.)*

We had access to two more data sets for the nuclear industry. The first was the industrial injury data for the United States, as published by the Nuclear Energy Institute (NEI) and the World Association of Nuclear Operators (WANO), two reputable industry associations. These data cover some 10 years of operation of more than 100 facilities throughout the United States, so they correspond to over 1000 years of accumulated plant operation.

In addition, we had access to a proprietary data set of SERs for a fleet of some 20 reactors of similar design covering some 15 years, representing another 300 years of operation and about 10,000 SERs of every kind.

It is useful to analyze these data with the DSM approach, to see how it compares, and to determine whether predictions can be made.

The result is shown in Figure 3.6 for the IR, where the accumulated experience is taken as calendar years for the United States, for which we have no data on the staffing levels year-by-year, but we note that the fleet of plants is roughly constant over that time. The data sources are the NEI and WANO.

Also shown in Figure 3.6 for comparison are the U.K. CIA data. We see immediately that the rates for both industries are remarkably similar in both magnitudes and trends, and that the minimum rate (IR) is about 0.2 to 0.5 per 100,000 h. This error

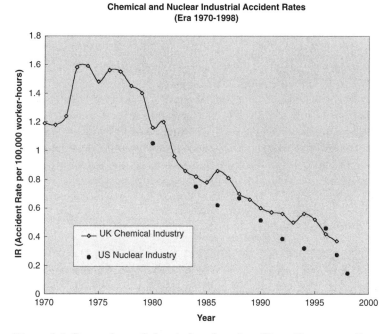

*Figure 3.6 Comparison of chemical and nuclear IR accident rates. (Data sources: U.K. CIA, 1998; WANO, 1999.)*

frequency value of 1 in 100,000 to 200,000 h is also close to the chemical and air-line error frequency values and represents our best choice for the minimum. The hazard value is within the statistical scatter and the 95% confidence intervals for all three industries.

We had not expected this numerical similarity between such apparently vastly different industries *a priori*. The fact that two independent industries in two countries can have nearly identical rates and learning curves was rather startling. It is a classic manifestation of the Universal Learning Curve at work in ACSIs, with an underlying human error contribution near the irreducible minimum. We presume the lowest observed value to be the minimum of ~200,000 h, at least as far as is achievable by society in modern technological systems with a high emphasis on safety.

*What the DSM has also told us is the strong emphasis on safety in the ACSI industries has netted about a factor of 4 to 10 reduction in the error rate, compared to the higher risk industries.*

### Criticality Accidents: An Analysis of the JCO Plant

So-called criticality accidents, where a sufficient mass of radioactive material forms a self-sustaining, heat-producing reaction by mistake, have occurred before, both in the early days of the nuclear weapons program and later in the development of nuclear fuels.

The strip in Figure 3.7 is the number of criticality incidents worldwide from the Los Alamos review for the years 1945–2000 and indicates a decline of incidents with experience. The number of such errors is small, partly because of the known hazard and partly because the plant and equipment are physically designed to limit the amount of material that can exist in any given place (typically a few kilograms). There are strict controls on inventories and on the storage of larger amounts, as well as the other types of materials (known as "moderators") that can be allowed nearby in case they help to promote the reaction. But criticality events still occur, and the known history (from Russia, Japan, the United States, and the United Kingdom) shows a small number of such incidents from about 1957 to 1965 as the technology was maturing, and isolated events occurring later, every 10 years or so.

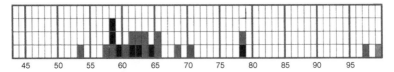

*Figure 3.7 Number of criticality incidents worldwide from the Los Alamos review for 1945–2000.*

In the Los Alamos report LA-13638, there is a review of many of these factors, with the emphasis on establishing cause. Some of the events had quite subtle *confluences of factors*, such as three people carrying a flask with sufficient liquid to form a critical mass when promoted (moderated) by their own bodies, thus killing them. All the events to date were due to human error: the latest one was at the JCO plant in Japan.

Let us look again at this JCO accident, the one with the "mixing using a bucket" incident. This is particularly interesting to examine, as there was a real event with serious consequences, and a *confluence of factors*. Can the DSM or any such approach help to tell if we could have predicted that, based on the operating history, something might happen? This is a simple but very challenging question, since prediction (forecasting) is not nearly as accurate as hindsight or Monday morning quarterbacking, which are totally accurate! But it is an instructive exercise.

We were kindly supplied with the data for the JCO plant by the Ministry of Trade and Industry (MITI) in Japan. Over the years of operation, all the industrial accidents were recorded, including quite trivial incidents such as a bee sting. So the database is quite sparse, because of both the limited number of employees and the small number of incidents, as shown in Table 3.2.

However, using the data given in the table, we can plot the incident rates. This is shown in Figure 3.8, where the accumulated experience is taken from the working hours by all the employees since 1981.

*Table 3.2 Incident Data for the JCO Plant*

| Calendar Year | Number of Employees | Total Number of Incidents |
|---|---|---|
| 1981 | 120 | 5 |
| 1982 | 147 | 4 |
| 1983 | 180 | 4 |
| 1984 | 179 | 5 |
| 1985 | 186 | 4 |
| 1986 | 178 | 7 |
| 1987 | 168 | 2 |
| 1988 | 163 | 2 |
| 1989 | 166 | 0 |
| 1990 | 168 | 3 |
| 1991 | 162 | 2 |
| 1992 | 163 | 3 |
| 1993 | 159 | 0 |
| 1994 | 149 | 2 |
| 1995 | 145 | 0 |
| 1996 | 131 | 0 |
| 1997 | 114 | 3 |
| 1998 | 110 | 0 |
| 1999 | 109 | 3 |

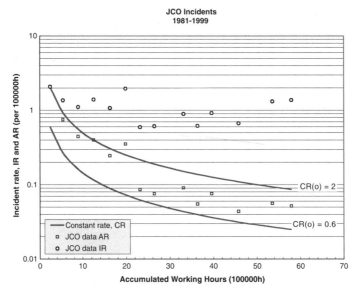

*Figure 3.8 The IR, AR, and CR for industrial incidents at the JCO plant. (Data source: MITI, 2000, author communication.)*

The data show a fair degree of scatter (in the time event history), but the average IR ~ 1 per 100,000 working hours. There is clear evidence of a rise in the rate after about 5 Mh, and no evidence of having reached a minimum rate. In fact the IR is rising slightly. This is shown more clearly by comparison of the AR to the CR, where the AR trends down. Two CR rates are shown, one corresponding to the minimum CR of ~0.6 (one in 150,000 hours) and the other to the initial (startup) value at the plant of ~2 (1 in 50,000 hours), which shows that a minimum of <0.6 per 100,000 h as obtained by the nuclear and chemical data had not been reached. Nevertheless, the IR is of similar order to those other two ACSIs.

We conclude that there is some hindsight evidence that the JCO plant had not reached the minimum rate, having departed from the learning curve, and also systematically increasing its incident rate after about 6 Mh of operation.

However, the overall rate is within a factor of 2 of the other industries, so examination of the yearly or average rate would definitely not have been sufficient to determine these trends.

This finding cuts across the typical divisions of regulatory authorities: those typically concerned with industrial (worker) safety and those concerned with nuclear licensing. It also means that short-term trends (i.e., yearly rates) are *not* good indicators and that longer term tracking of trends is key.

### *Safety Trips and Lost Production: Choosing Measures for Accumulated Experience*

Trends can be followed in many forms and for many chosen parameters, so it is worth examining a key area of industrial importance: namely, lost production due to process problems or plant unavailability.

Safe operation is key, and safety systems are included in production and operating facilities specifically to ensure safe operation. For example, like all technological systems, chemical plants are equipped with safety instruments and controls, designed to keep the system operation within safe limits (in pressure, flow, temperature, etc.). In a simple analogy they are like the home thermostat, turning the process on and off as needed. If the preset limits or some agreed conditions are exceeded, the plant is shut down—"tripped off line" in the terminology of the industry. The trip can be initiated by the process control or by safety systems, which may be independent.

For production and manufacturing facilities, meeting quotas or orders of a needed product for a customer is the normal practice, be it fertilizer, paint, tools, pressings, parts, or chemicals. Trips mean lost production time and product, so it is a good idea—or plant and management goal—to keep the trips to a safe minimum. If trips are occurring and moving the trip levels is not safe, then only smarter operation and/or greater margins can reduce the trip rate. The U.S. commercial nuclear power industry has specific goals for the number of trips for a plant in a given year based on the principle that good operation is safe operation. How do we know if we are learning to better produce?

Trips are usually recorded on a calendar time basis (e.g., per year). We were fortunate to be given access to some detailed data for a prototype production plant, which was just going through the startup and early demonstration production phase. The plant, process, and technology were all new. The operating staff had to learn to operate the plant, trying to maximize the plant output and availability, and thus demonstrate that the plant could keep the product cost down.

The standard trip rate reported for the plant operation is shown in the top of Figure 3.9 as a rate per operating hour (taking out shutdown or nonoperating time) versus actual operating hours. There is no real learning pattern, and trips per operating hour are actually increasing with operating time. But when converted to the IR as the rate per unit production and plotted against the accumulated experience of the accumulated product, a clear learning curve emerges. As shown by the lower plot, the trip rate is actually decreasing with increases in accumulated product and is following the exponential learning model.

We all can see how the two curves apparently behave completely differently based on the choice of the accumulated experience explanatory variable. The clear implication of this simple analysis and demonstration of the typical DSM approach is that the operating and management staff were not just learning how to operate a large and complex

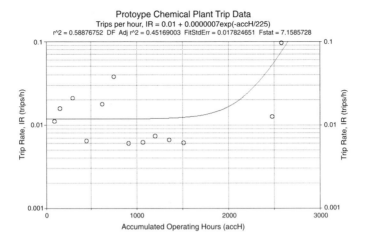

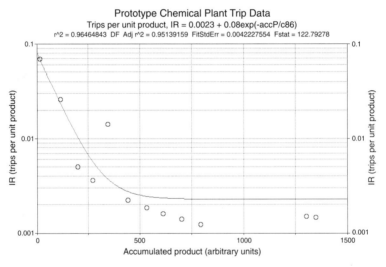

*Figure 3.9 Demonstration of existence of a learning curve in chemical product manufacture, with trip rates plotted versus operating time (top) and versus accumulated production.*

piece of new equipment. They were *actually learning how to make more products*, while still maintaining a safe working environment and equipment.

A similar trend of decreasing errors with increasing productivity (and vice versa) is evident in the data from the U.S. underground coal mining industry from 1931 onward.

As shown in Figure 3.10, where the data are plotted, a learning curve is evident in the injury rate when we choose production of coal in tons per hour worked as the accumulated experience basis. The loops in the curves indicate hysteresis, so that when production falls the rate has followed nearly the same trend backward. What this analysis

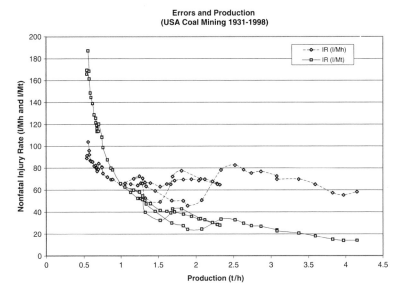

*Figure 3.10 Evidence of technological learning curve with increasing productivity. (Data source: U.S. MSHA, 2001.)*

shows is the importance of the choice of the "right" measure for the accumulated experience when examining or determining technological or production learning curves.

## 3.5 HUMAN ERROR AT WORK

No work is quite error-free, it seems, and wherever we looked we found this to be the case. We stumbled across the U.K. Government's Expenditure Plan for the vast Department of Trade and Industries (DTI) for the fiscal years 2001 and beyond; this unlikely source contained key error information for purely paper processing.

One of the bureaucratic but necessary tasks of the DTI's Imports Licensing Branch is to process and issue some 30,000 to 40,000 import licenses a year, with the twin challenging and potentially conflicting goals of processing each license within 5 working days and of reducing errors. The reported error rate for import licenses has steadily fallen since 1996 and follows a learning curve when the accumulated experience is measured by summing the number of licenses processed. The reported error rate decreased to about 0.2 errors per 1000 licenses by 2001, or 1 error per 5000 licenses (or ~2 × 10$^{-4}$ per license). Assuming the 5 working days it takes to issue an import license is typical for all licenses, that is an error interval of

= 5 days times 8 working hours per day divided by the error rate of 0.2 per 1000

= (5 × 8)/0.0002 hours

= 1 error in 200,000 processing hours

We have repeatedly found this order for the minimum error interval before. This unexpected and new occurrence clearly indicates that *errors at work in a modern office environment also occur at about the same rate as in all other technological systems.*

Errors of this form can manifest themselves as inconveniences and hassles in many ways. Anyone who travels on business by air knows that checked luggage gets lost sometimes. The data for lost baggage claims on U.S. airlines for 1990–1998 still show no significant reduction from a rate of a few per 1000 passengers, despite the high cost and hassle factor.

We have so far assumed that the minimum error value, whatever it is, represents that irreducible piece of the error rate that is due to human error, which is present as an underlying contribution throughout the learning process. The basic and underlying minimum error frequency deduced so far is one in approximately every 200,000 of the working or accumulated hours of experience, which is roughly every 20 years. Be it licensing or baggage handling errors, we are exposed to that underlying risk of all-too-human mistakes.

Various operational activities have an estimated human error fraction or contribution to accidents and errors of order 60% or more (e.g., in auto accidents, ship losses, and airline crashes). Thus, human error is the major contributor; we wondered what the fraction might be for the almost completely safe industries (the ACSIs).

This question leads us to a diversion into the field of endeavor known as "human reliability." There is a vast body of work trying to define and determine the human error probability (the HEP as it is called), using human reliability analysis (HRA for short).

The HEP is analyzed by watching humans perform simple and complex tasks, often on simulators to mimic the actual interface with the technological system, such as the plant, equipment, control room, and working conditions. The observations of the human performance include the effects of individual tasks and actions, team and working environment, stress and time pressures (so-called performance-shaping factors or PSFs), and organizational factors. These last organizational elements include management attitudes and situational and context effects, as well as safety management systems.

The output of the work for practical uses is often a numerical estimate of the HEP (including the PSFs) as a function of the fault, incident, or event duration as it unfolds.

This whole lexicon of terms and abbreviations itself can be as confusing as the errors themselves, and the studies, such as those of Reason and others, delve deeply into physical, psychological, and technological factors. In the imprecise science of understanding the workings of the human mind and psyche, there seem to be as many "models" for what and how many of these factors exert their influence as there are authors and researchers. The subject of "human reliability" is still evolving, and much valuable information has been gathered.

Once again, we will avoid preferences or choices for any of the many other theories. We will appeal only to the error data and what they tell us, and to the simplest model of all, the Universal Learning Curve. In our DSM, we think of the HEP as equivalent to the error rate (A* or E*); the context and PSFs then determine the slope of the (exponential) learning curve (the rate of learning); and the time evolution of the whole event is effectively the accumulated experience. The MERE describes and quantifies the underlying HEP component of the Universal Learning Curve for the given human–technological system.

Whereas the HEP is derived by decomposition of the probability of error for a series of differential actions or their elements, the DSM provides an integrated or overall error rate without consideration of the individual task details. In other words, we are dealing with the errors in the overall technological system, not in the bits and pieces that it is made of.

We have other information and data that bear on the event and error rates, from the SERs. These are quite exhaustive data files that break down the significant and reportable events by system, plant, year, cause, and type. As noted above, the files we examined had data on more than 10,000 such events over many years; the ISPN also filed more than 900 in a year. Many of the events (SERs) do not constitute a real operational hazard, but are a reportable departure from the expected or licensed operating state or condition with the potential for a safety implication. We focused on the contribution of human error to the SERs and analyzed the fraction of the total SERs that were classified as being caused by human error (HESERs). We separated these events out and plotted the ratio of HESERs to the total number of SERs as an IR, with the result shown in Figure 3.11.

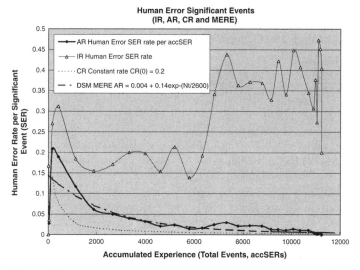

*Figure 3.11 Typical set of nuclear industry significant events, SER, showing the IR and AR fraction classified as due to human error.*

The basis for the accumulated experience is the total number of SERs, since each plant started in a different year. The intercomparison between them cannot make sense on a calendar-year basis if they started at different times, and all have different histories. This choice also tests the notion of whether there is learning actually occurring from the SER reporting and root cause analysis process and what the fractional human error contribution might be.

The fraction due to human error is apparently ~20–35%, and the IR varies with the accumulated experience. As is evident, the fractional HESER IR fluctuates and rises beyond about total 6000 SERs, with the fluctuations at the very end due to a small data sample (for SERs > 10,000). Therefore, there is evidence of learning (the decrease in the fraction) and then of forgetting (the rise after about 6000 SERs), thus forming a classic "bathtub" curve with a dip in the middle.

The fit derived from the DSM is also shown in Figure 3.13 for the AR, and the best-fit MERE line from TableCurve is

$$A = A_M + A_0 \exp(-N^*)$$

or

$$AR = 0.004 + 0.14 \exp(-accSER/2600)$$

The results imply there is indeed a lower bound, below which the fraction of HESERs does not fall, and the constant rate CR line for $CR(0) \sim 0.2$ is therefore shown for comparison. This DSM analysis indicates that more than 20% of this set of SERs are due to a significant contribution from "human error" and show systematic trends with accumulated experience.

### Recovery Actions and Human Error Rates

Perhaps the other most significant item to emerge from the analysis of large and rare accidents is the realization of the importance of so-called recovery actions by the operators of the technological system (be it pilots, drivers, or operators). So how good are we at recovery actions when tackling complex situations?

These actions can include correcting previous errors, fixing the systems, or finding unknown or hidden faults. The recovery actions may be outside of normal experience. They rely on ingenuity, judgment, and training, as well as acute situational awareness, and the availability and interpretation of good data or information on what is happening. The actions may involve using the available equipment in new and different ways, and truly taking charge rather than just letting the technological system evolve.

This is the core of the human–machine interface and the heart of "human factors" as a science, rather than as pure ergonomics or workplace and machine design, even though these are important, too.

The RECUPARARE method at IPSN carefully classifies the French SERs (errors). More specifically and importantly, more than just tracking what went on, the IPSN approach analyzes the recovery of the situation based on the human system and/or safeguarding of automatic systems. Thus, so-called latent faults are recovered or corrected after activation of the technological system, as well as faults appearing when it is already in service.

A fault is considered "latent" if it is present in a system before activation of the system without being detected. Latent faults are especially important, as the safety of facilities can only be ensured if all of the technological systems (and the safety systems) are actually available, that is, if they do not have latent faults.

So *not* detecting a latent fault (which is one of our *confluence of factors*) is a failure or error probability that can evolve and change with time. It is less likely to be immediately found but more likely to be corrected after being diagnosed as present, having an impact, or simply being found. In the DSM terminology, we would say it is more likely that the problems and errors are found as experience is accumulated and learning occurs, in this case as the incident itself evolves in real time.

Many things may affect whether the "correct" actions are taken to recover a situation in the physical and mental environment. This can include factors such as time, training, mental models, procedures, familiarity, teamwork, tiredness, stress, personality, lighting, instruments, controls, and alarms. This is generally known as the "context" of the event, and hence, actions, events, or errors should be viewed in the context in which they actually happened rather than as single or multiple operations. In the terminology that we have used until now, this is the "*confluence of factors*" that we refer to, and how fast the "learning curve" for the actual event and set of errors is being followed.

The totally independent studies of Baumont *et al.* (2000) of the French data support this and led them to state:

> Analysis showed that incidents regarded as the most significant for safety by the IPSN analysts—due to their human context and potential consequences on safety—are incidents for which detection or recovery is particularly lengthy. This result shows that detection and recovery times are probably pertinent indicators of safety performance.

The IPSN graphs of the failure probability to detect latent faults (the human error probability, p) for 2 years of data were kindly made available to us, as shown in Figure 3.12.

As can be seen, early on in the fault or event, with event times of less than, say, 20–30 minutes, there is an error probability of 90%. The error remains: there is just not enough time to correct it. From about 1000 minutes (15 hours) there is about a 50:50 chance of fault detection (error correction) and recovery action. By 20,000 minutes (nearly 2 weeks) there is about a 70% chance of recovery, or about a 1 in 3 chance that the fault has not been found.

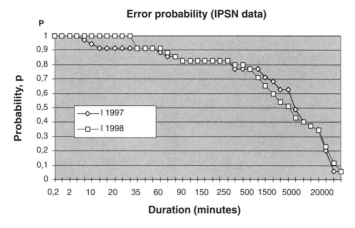

*Figure 3.12  The probability of nondetection of a latent error as a function of the event time for the two years 1997 and 1998 of data. (Source: Baumont et al., 2000, analysis of EDF operating data.)*

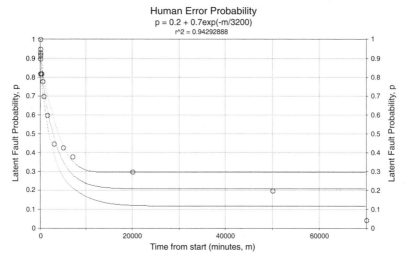

*Figure 3.13  DSM plot and MERE curve for the IPSN data shown in Figure 3.12, with the dotted lines showing the 95% confidence levels.*

We believe this decreasing error probability with elapsed time literally shows learning at work, in real faults. It is difficult to read the data (numbers) from the graph, and we were kindly supplied with good electronic file copies by the group at IPSN. Reading the data on the "tail" of the curve with the odd scale was still difficult and subject to some error. They were unable to provide us with the actual numerical data. We would expect the data to follow an exponential learning curve. To test that hypothesis, we replotted the information on a linear accumulated-experience basis to see if the data were indeed following a DSM exponential learning curve. They do (Figure 3.13).

As shown in Figure 3.13, the data actually follow a learning curve, and the simple MERE exponential model curve fit of the probability, p, for these data is then

$$A = A_M + A_0 \exp(-N^*)$$

i.e.,

$$p(IR) = 0.2 + 0.7 \exp(-m/3200)$$

Figure 3.13 shows that the data are very sparse at the later times, and the minimum (the tail of the curve) has only two or three data points at very long times. The overall fit is based on all the data, so the fit is not too sensitive to the value. The asymptotic lowest probability recorded in these data is in the range 0.1 to 0.3 at about 70,000 minutes. This is a lower minimum error frequency of 1 in $70,000/(60 \times 0.1)$, or 1 in ~17,000 hours as a *prediction* from using the DSM. The lowest actual probability data point at long times is ~0.045, which gives 1 in 26,000 hours.

Given all the uncertainties in this type of analysis, the best we can say is that this implied human-error result is somewhat higher than the minimum error frequency range that we have found elsewhere (of 1 in 100,000–200,000 hours), but identical to that for industrial errors (of order ~25,000 hours). In other words, real everyday events apparently show a minimum human error rate in real time that is larger than the minimum attained or shown by errors in large historical hazard data sets. To our knowledge this fact has not been shown before.

The problem is that data collection and analysis on events and errors must go on for very long times and must also examine many, many events in order to obtain accurate statistics for making reliable estimates of the minimum human error probability.

*We can conclude that actual experimental data, and these real data on actual significant events, do indeed show there is a significant human error contribution to the minimum rate.*

### The Impact of HMI Technology Changes on Event Rates

Can this minimum error rate or contribution be reduced by changing the technological interface?

One might intuitively expect changes and technological advances to the human–machine design or interface, or the HMI as it is called, to have positive impact on the learning curve. For all technological systems, there has been much work on trying to improve the control layout so that errors are minimized, data and status are well presented and clear, malfunctions are easily recognized and handled, and the environment around the interface is conducive to good decision making and work habits.

Increases in automation have occurred, both to lessen the workload and to improve the control characteristics, as well as to more readily assimilate and present the system status to the human watching over it. This also takes advantage of modern computer systems and improvements in the graphic user interface (GUI), and modern technologies that include computerized displays and "intelligent" processing of signals and alarms. The computerization of procedures is also possible under the general heading of computer-aided operations and safety systems (COSS).

The aircraft industry has been at the forefront of this effort, because so many fatal accidents occur on takeoff and landing (about 80%), which is only 5% of the flying time. However, it is the high-stress time, with information, workload, stress, and traffic all at their maximum. So automation of many aspects of flight has occurred, and the cockpit has become more computerized than in the purely mechanical days of old.

The major manufacturers (Boeing and Airbus) automate to different degrees, with the tendency for more automated operational controls in the later A300 series aircraft than the later B700 series aircraft.

To examine the question of what insights the DSM can provide here, we were kindly supplied with some proprietary data on events in an operating fleet of real commercial aircraft. These were flight operation event data (FOQA) for two types of aircraft cockpit (which we will designate generically as Type A and Type B). These two types were based on different cockpit technologies *for the same plane design*, to see what impact the technology change might have on event rates and learning curves. Type B is the "old" cockpit style, with gauges and dials and less automated analysis, and is the control population. Type A is the "new" style, with more so-called "glass" displays, which are computer screens that digest and present information more compactly and rely on greater computerization.

The basis for the accumulated experience was then the actual accumulated flights for each aircraft cockpit "type," which resulted in about 150,000 flights for both together. The results are shown in Figures 3.14 and 3.15 for both the IR and AR. The trends clearly indicate that one technology type has a systematically higher event rate than the other, but that the learning rates (slopes) are similar. The "old" technology always has a lower rate of events than the "new."

For Type A (new), this CR line is based on an initial event rate of 33.5 per 10,000, and for Type B (old), 8.5 per 10,000. These lines fit the trends reasonably and show that in both cases the trend is toward a constant rate, which is systematically different for both types. Also, the Type A shows a slight increase in rate over the last 20,000 flights, as evidenced by the upturn in the curve. Looking again at Figure 3.14, we can see this increase there, too, although it is somewhat masked by the general ups and downs in

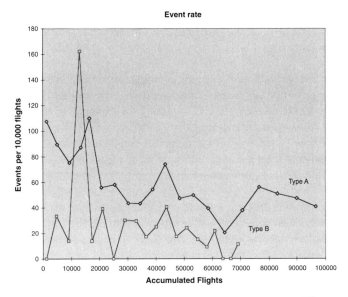

*Figure 3.14 IR for events with one aircraft type with two different cockpit technologies.*

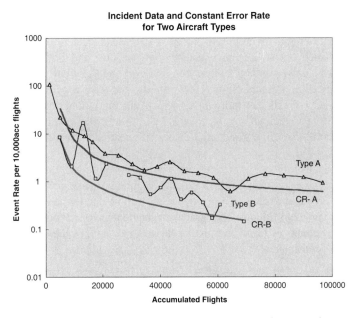

*Figure 3.15 AR and CR for events with one aircraft type with two different cockpit technologies.*

the IR. The minimum rates from Figure 3.15 seem to be ~0.1 and 1 per 10,000 accumulated flights for Type A and B, respectively. This would appear to be a statistically significant difference.

Not only is this a very subtle and apparently contradictory result, but we have also shown that the human error rate is *dynamic*, in the sense that it varies with time and is altering as learning occurs. Models for HEP are used, for example, in probabilistic safety analyses (PSAs) using event and fault trees coupled with HEP curves. Analyses of accident sequences where recovery actions are involved are unlikely to accurately reflect improvements from learning unless the dynamic changes in the error rate and the minimum rate are accurately modeled.

What we have shown is that our working environment—where we may spend a large part of our lives to make a living—is also subject to errors. The human contribution is significant and often varies with time. Technological changes intended to improve the working environment, and perhaps intended to reduce errors, should be carefully tracked to determine their actual impact on error rates. Almost completely safe industries have error rates of order 1 in about 100,000 to 200,000 hours, rates that are completely consistent with the lowest or minimum rate that we have found elsewhere. This value is explicable on the basis that it is due to the common thread of human involvement.

### Failures and Defects in Pressure Vessels

The Industrial Revolution of the past 200 years was based on the harnessing of steam and other energy sources to drive factories, machinery, plants, and equipment, using the mighty steam-engine inventions of James Watt and his great 19th-century contemporaries. Nowadays, the steam can be used to drive turbines and to make electricity as well. The steam is raised in large boilers and heat exchangers, and the water and steam can be at a very high pressure (some 1000 pounds) and temperature (many hundreds of degrees). Such vessels can fail, sometimes spectacularly and with loss of life, so pressure vessels (PVs) are an area of great industrial safety significance, where insurers are involved in inspections and strict construction standards are also imposed.

Despite the adoption of best practices and rigorous requirements for materials and safety margins (e.g., as in the ASME Codes and Standards and proposed new European rules), PV failures and defects still occur. Failure can be due to many causes: in-built defects, operating errors, in-service degradation, inadequate inspection, manufacturing errors, material cracks or imperfections, or just poor design, or these factors in some combination or confluence.

Bull, Hoffman, and Ott (1984) studied marine boiler history, focusing on main boiler failures in ships from 1882 to 1974. They researched about 800 failures and observed a decreasing failure rate with year of failure (a learning curve) from initially about 1 per 250 ship-years down to 1 per 2500 ship-years. They stated that "human error as a cause of failure shows no significant decline over the whole period." For more recent boilers,

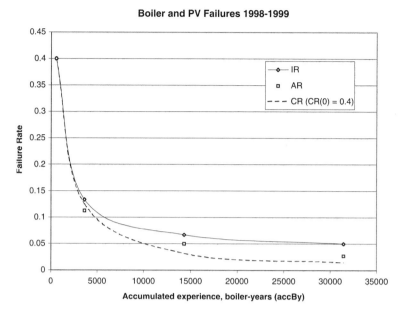

*Figure 3.16 Boiler failures and defects for all types show a clear learning curve. (Data source: U.K. Safety Assessment Federation, private communication, 2001.)*

their analysis showed no decline in failures for main boilers and a small decline for auxiliary boilers.

We wanted to reanalyze the errors for this classic case of PV failures, but such data are quite hard to find. We were kindly given access to the latest and most systematic modern data files by the U.K. Safety Assessment Federation, which listed failures and defects for 1998 and 1999. The data were for water tube, shell, and PV boilers, most failures and defects being for the last two types. There were nearly 2200 failures, with the data grouped by age of the boiler in increments of 5 years and by failure type. As a first pass, we used the same DSM approach we had used for Berman's shipping losses. From the age distribution of the failures we found the boiler-years to failure and calculated the IR and AR. There were about 31,000 accumulated boiler-years of operation in 1998 and 1999, which we took as the working basis for the accumulated experience. The result of this calculation for all failures and defects for all types is shown in Figure 3.16.

The DSM shows a clear learning curve, but the rate is not declining as fast as the CR based on the initial rate of ~0.4 per boiler-year (By). This confirms the prior observation that the failures are not declining fast enough.

Now the MERE exponential model for these data is given by

$$\text{IR (failures per By)} = 0.05 + 0.455 \exp(-\text{accBy}/2133)$$

The *minimum failure rate* is then given by 0.05 per By, and the failure interval is

$$= (365 \times 24)/(0.05) \text{ per boiler-hour}$$
$$= 8760/0.05 \text{ per hour}$$
$$\sim 1 \text{ per } 175{,}000 \text{ hours}$$

This is very close to the other estimates that we have for minimum failure intervals, which are typically 1 per 200,000 hours. *We conclude that this important technological system exhibits precisely the same confluence of (human) errors as all the others that we have studied.*

So we humans embed our errors in the systems we design, manufacture, and operate— and these errors surface in the events and accidents that result.

# 4

# LIVING IN SAFETY

*"Life is safer than it has ever been, but we are no longer prepared to accept any risk in anything we do."*

—Alice Thomson, *Daily Telegraph*

## 4.1 RISKY CHOICES

We exist in a technological society, and risk is all around us.

We would all like to live safely, going on with our normal lives, without fear or risk. But we have the added hazards of traveling and going to work, as well as the risks of working in our modern technological society. Just the act of existing and everyday living exposes us to risk. Breyer's study (*Breaking the Vicious Circle*) examined the regulation of risk, how it is assessed in the United States today, and how it must be improved. He methodically defined four activities:

1. Identifying the potential hazard or risk
2. Defining the harm from the risk exposure
3. Estimating the amount of actual exposure
4. Categorizing the result (i.e., as harmful or not)

Breyer goes into the details about how to prioritize what risks should be worked on, and how to define a real risk, especially for the chemicals and carcinogens we are all exposed to. He showed that we can and do spend vast sums of money per life saved on some very low risks (on substances such as toxic waste, asbestos, and Superfund sites) and very little on many real ones (on topics such as improved diets, reduced sun-bathing, early cancer screening, and subsidized smoke detectors).

However, as Breyer and others discuss, our attitudes to the risk are colored by whether we think or perceive the risk as unknown and/or hold it in fear or dread. The more it is unknown and dreaded, the more we desire the risk to be regulated or personally avoided. *Thus, the public's (yours and mine) perception of risk is highly colored not by the real risk, but by the perceived risk.*

For instance, nuclear power is perceived as a risk, whereas prescription drugs and antibiotics are not. We choose to go to the doctor for a prescription, placing our life in the physician's hands, literally, as a voluntary risk. We perceive the benefits (of getting well) as far outweighing the risks (of disease and death). After all, our doctor is qualified, trusted, and needed. We have all heard stories of poor treatment or operations that failed or were on the wrong patient. But how many deaths are there from medical errors? Will more people die from the errors that occur in modern technological medicine and in preventative and diagnostic care? *Are you safe?* There is growing public concern over medical standards being achieved. Reviews of best practices are being undertaken in many countries, concerning improved monitoring and auditing of standards, as is the common practice for other high-tech industries. Confidential reporting schemes are under discussion.

## 4.2 MEDICAL SAFETY: MISADMINISTRATION, MALPRACTICE, DRUGS, AND OPERATIONS

The medical profession has only recently begun to study its rate of errors, and the study of medical errors has been very active with the growing realization that there is a real problem. In the terminology of the medical profession, these are called "adverse events," including errors, deviations, and accidents (AMA National Patient Safety Foundation). The general field of study of medical error now has the almost euphemistic term "patient safety"—but it is not usually the patient who is making the error. It is also often implicitly assumed, in lawsuits and the like, that it is the medical doctor who is responsible for patient safety when in fact it may be the medical system.

The medical industry and insurers, as well as the profession, are also apparently deeply concerned about errors. Medicine is both big business and big risk, and the term "quality assurance" is also sometimes loosely used in relation to this general area. A recent industry initiative called the "Quality Care Research Fund" places its priority on strategies to improve patient safety (i.e., to reduce deaths) and to reduce the unacceptable number of avoidable medical errors (i.e., to reduce insurance and liability exposure).

Medical errors are newsworthy:

> "Wrong doses kill 1,500 each year," said one headline, because heart patients received the wrong doses of drugs some 5–12% of the time (roughly 1 in 20 to 1 in 10 times).

"Medical mistakes kill 98,000 Americans a year" ran the stories that accompanied the release of the National Academy's Institute of Medicine (IOM 2000) report *To Err Is Human*.

"Deadly Prescriptions: adverse drug reactions kill an estimated 10,000 Canadians a year" was the bold headline in the Observer Section B of the Ottawa Citizen of April 7, 2001.

The practice of medicine is now a complex modern technological system. As in all the other fields we have studied, there are many types of, and opportunities for, medical errors in the system:

- Misdiagnoses of the patient's condition
- Misadministrations of drugs and treatments
- Mistreatments of the patient
- Professional negligence
- Mislabeling of drugs and equipment
- Incompetent treatment
- Patient error
- Malpractice
- Miscommunication

Modern medical care relies to some extent on the technology afforded by highly refined diagnostic tools, such as magnetic resonance imaging (MRI) and computerized tomography (CT), and on prescription medications, while still retaining a degree of dependence on the art of clinical acumen and diagnostic intuition. Medication errors can occur in the prescribing, dispensing, administering, monitoring, and control of drugs. The prescription may itself be poorly written, illegible, or misread.

There is much sensitivity in the medical profession to the admission of error and the assignment of blame. The increase in litigation follows the legal profession's introduction of "contingency fees" (fees that are only paid when a claim is judged), which encourages aggrieved parties and lawyers to sue for financial gain as well as recompense. The consequences of errors—aside from the "adverse event" to the patient—are potentially large in terms of legal, product, and fiscal liability, insurance premiums, and possible personal and professional disgrace. Compared to other areas, we found it extremely difficult to obtain real data for errors, despite there being so many of them. A search of available websites revealed nothing in any technical detail. Confidentiality is a tradition highly maintained, but the public is demanding change. Our efforts to talk more and find out about more error data from the IOM authors and others failed or were politely rebuffed. They were basically "too busy"; the railways and other industries had much more safety information that was far more easily available.

The publicly available information on medical errors that we could find was embodied in the references to the IOM report: everything else was a blind alley. We therefore obtained all the referenced papers containing the error studies and analyzed these using the DSM.

The IOM defined an error as the failure of a planned action to be completed as intended (execution error) or the use of a wrong plan to achieve an aim (planning error). An "adverse event" was then defined as injury caused by medical management rather than by the patient's underlying condition.

There is still debate over the exact error numbers, whether some people would have died anyway, the sample sizes used in the studies, and the accuracy of the reporting and classification schemes. We avoid these professional, liability, confidentiality, and fiscal debates, go directly to the data, and ask what are the error rates and what do the data tell us? How does the error rate compare to others? The results are interesting to all those with a personal stake in medical care.

### Medical Errors

So how safe is your trip to the doctor or the hospital?

The death rate for medication-error deaths, D, from 1983 to 1993, for the United States, has been studied using an international classification scheme. These deaths were defined as accidental poisoning by drugs, medication, and biological causes as a result of recognized errors. The data may include some nonmedical causes (e.g., addiction) since the analysis used the death certificates (cause of death) for a population of some 220 million people.

For the measure for the accumulated experience, we took the total population, which was given, thus correcting for the increasing numbers and avoiding the calendar-year reporting problem. If the number of patients was less, it is presumably some fraction of the total population. The result is shown in Figure 4.1, where the number of deaths per million (D/M) are shown as the IR and CR. It can be seen that the IR rises with increasing experience (a sign of insufficient learning), and this is confirmed by the trend and comparison of the AR with the CR line on the graph of about 12/accM. (This rate is intriguingly close to the commercial aircraft fatal accident rate of ~15/MF, but is not comparable yet, as we shall see.)

To show the trend in more detail, Figure 4.2 has a number of CR lines drawn as a network where the actual data cross the lines. The AR MERE is given by analyzing the trends in the Phillips data (Figure 4.3), which gives:

$$A = A_M + A_0 \exp(-N^*)$$
$$\text{AR (deaths per accM) D/accM} = 2.6 + 16.5 \exp(-\text{accM}/392)$$

This analysis implies that the minimum AR will be ~2.6/accM and hence deaths will rise inexorably with population and experience increases. With an accumulated

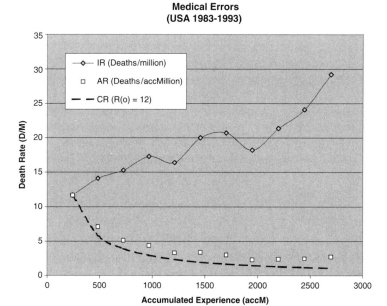

*Figure 4.1  The IR, AR, and CR trends in the United States for deaths due to medication errors. (Source: Phillips et al., 1998.)*

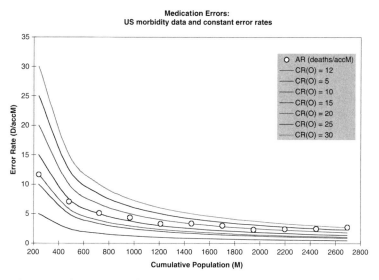

*Figure 4.2  Comparison of mortality error data to constant rate lines.*

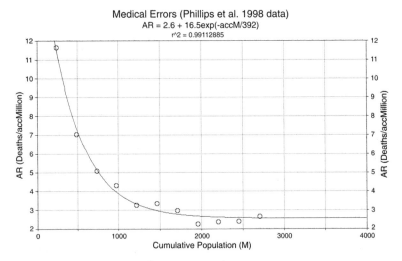

**Figure 4.3  The U.S. mortality data plot for medical errors using the total population as a basis for the accumulated experience.**

population in 2000 of ~3400 million people since 1983, that irreducible minimum error (death) rate is

$$= 2.6 \times 3400 \sim 8800 \text{ per year.}$$

This is the *predicted* underlying or asymptotic error rate.

### Medical Error Sample Size and Experience

All other medical error studies are for smaller populations, using sample studies for particular hospitals, regions, and/or years. They are or have been used to extrapolate to larger populations (e.g., national trends), and that is how the IOM estimates were made. We summarize the data in Table 4.1.

The sample sizes (populations studied) are very different, ranging from ~200 million to a few thousand in clinical groups, just like the problem we studied with tire failures. The apparent error rates are very different, ranging from 10 to 150,000 D/M (i.e., from 0.001 to 15%). How can we plot these on the same graph? Are the death rates really that different?

These apparent rate variations are, we believe, an artifact of the differing sample sizes, just as was the case for tire failures. This sample or observed population size variation affects the observed error rate via the large variations in the accumulated experience (compare also the airline case).

To compare error rates, we must determine what is the appropriate accumulated experience basis (the accumulated number of patients). For the DSM, we treat the rates as

*Table 4.1  Summary of Medical Error Studies Reported in IOM, 2000\**

| Study | Date | Number of Errors | Deaths/1,000,000 | Sample Total |
|---|---|---|---|---|
| Phillips *et al.*, 1998 | 1983 | 2800 | 11.67 | 240,000,000 |
| | 1984 | 3400 | 14.11 | 241,000,000 |
| | 1985 | 3700 | 15.29 | 242,000,000 |
| | 1986 | 4200 | 17.28 | 243,000,000 |
| | 1987 | 4000 | 16.39 | 244,000,000 |
| | 1988 | 4900 | 20.00 | 245,000,000 |
| | 1989 | 5100 | 20.73 | 246,000,000 |
| | 1990 | 4500 | 18.22 | 247,000,000 |
| | 1991 | 5300 | 21.37 | 248,000,000 |
| | 1992 | 6000 | 24.10 | 249,000,000 |
| | 1993 | 7300 | 29.20 | 250,000,000 |
| Cohen *et al.*, 1986 | 1975–78 | 2634 | 91,980.00 | 26,003 |
| | 1979–83 | 3383 | 98,460.00 | 30,979 |
| Brennan *et al.*, 1991 | | | | 30,195 |
| | | 98,609 | 36,906.46 | 2,671,863 |
| CMA, 1977 | 1977 | 870 | 41,698.62 | 20,864 |
| Barker and McConnell, 1961 | 1961 | 93 | 162,587.41 | 572 |
| Barker *et al.*, 1966 | 1966 | 2920 | 265,093.05 | 11,015 |
| | | 393 | 129,148.87 | 3043 |
| | | 1461 | 149,249.16 | 9789 |
| Shultz *et al.*, 1973 | 1973 | 196 | 53,289.83 | 3678 |
| | | 22 | 6,382.36 | 3447 |

\*References as given in and by IOM report.

the AR, where each study uses and corresponds to a different accumulated experience. The result is shown in Figure 4.4, where the constant rate line, CR, is ~50,000 deaths per million (D/M) patients due to medical errors (i.e., a risk of patient death of about 5%).

There is much scatter in the data, and it is always possible that we have made an error! Since we estimated that the irreducible minimum error rate as ~8000/accM, the DSM analysis supports the need for more learning, as did the IOM 2000 study. It also suggests that the present, nearly constant rate can be reduced by about a factor of 5 or so, from the present 50,000/M to ~8000/M in the future. This reduction potential is consistent with what we have found for the reduction in error rates possible by ACSIs, which practice extreme safety management techniques.

At the least, the DSM has shown a way to compare and contrast the apparently very different results of the disparate error studies obtained using various screening criteria, sample sizes, and populations. *Also, once again we have shown an underlying constant error rate value, as we did at the very beginning of this text for aircraft.*

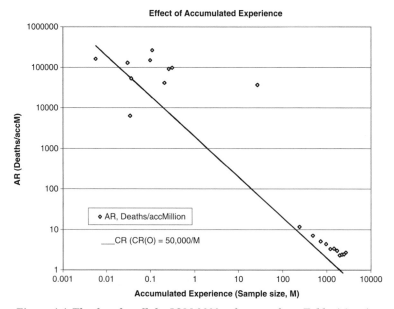

*Figure 4.4 The data for all the IOM 2000 references from Table 4.1, using the accumulated experience based on sample size: the straight line is the CR of ~50,000/M.*

To turn the error rate into frequency, we need to know the actual actions and work-load (i.e., the technological system error rate). We have measured the outcome of the error (as death) of the patient but not the cause. Assuming a medical system had an error rate of 8000/accM, consider the error frequency if a system was treating P patients a day, during 8 hours of work. But we know that the actual patient only sees *active* treatment for very little of that time, T, say, perhaps for 10 to 15 minutes of the day and could have multiple treatments, D, say (e.g., decision points, interactions, medication doses, X-rays, and dressing changes). Then the expected error frequency can be written as:

$$= (8000/1,000,000) \times (D \times P \times T)$$

Unfortunately, we do not know the values for D, P, and T to put in here. So we try another approach: assume an "equivalent" single system with an error rate of 8 per 1000 patients. This fictitious medical system sees 20 patients in each day; say 20 patient-days or about 200 patient-hours. The error interval is then

$$= 1 \text{ in } 200 \times (1000/8) \text{ hours}$$
$$= 1 \text{ in } 25,000 \text{ hours}$$

or about once in three years, the same error interval as industrial injuries.

Recall that a *confluence of factors* is also often involved: a mixup in drugs or their labeling, or the wrong dose is administered or prescribed, or the wrong drug is prescribed, or it is given to the wrong patient.

The medical system consists of physicians, nursing staff, technicians, testing labs, equipment manufacturers, drug makers, and management, all overlooked by regulators. Insurers in the United States foot the treatment bill and separately cover liability. High quality is expected and needed in the system. But is the error rate high compared to what we might have expected from a technological system anyway?

### Radiation Therapy Errors

There is another source of data on medical system errors. Over the past 50 years, the great medical advance of radiotherapy has emerged. By using various types of radiation and radiation sources, the treatment aims to kill or shrink malignant tumors, alleviate pain, and help make diagnoses. The radiation type, duration, and magnitude are calculated and estimated, and different doses of radiation used for different locations. There are several million treatments or "procedures" a year in the United States alone, saving and extending countless lives. The benefits of successful treatment are unquestioned, and the treatment is usually intended to help and not harm the patient. It is a fine example of a truly modern technology applied to and for the benefit of humans.

But mistakes do happen: humans must estimate and administer the treatments to patients using machines and/or medical equipment. Mistakes can occur throughout the radiotherapy treatment process, since treatment involves many types of radiation "sources" and different radiation doses for the many different parts of the body. The operators of treatment machines and radiation sources are licensed by the U.S. Nuclear Regulatory Commission (NRC), and hence data on misadministrations are available because the licenses require errors to be reportable.

We were fortunate to be able to examine some specialized and rare data available from the licensing of radiation sources for the treatment of patients with tumors and cancers. This therapy is designed to administer the right amount of radiation to kill malignant cells and to relieve pain, in single and/or multiple doses. The treatment is highly beneficial, is routine, and saves many lives. The application of such treatment is highly skilled and potentially dangerous, which is precisely why the use of the radiation sources is licensed by the U.S. Nuclear Regulatory Commission (U.S. NRC).

Multiple human errors were cited by the NRC as responsible for about two-thirds of the misadministration (wrong dose) cases, completely in line with the human contribution found in our other studies. The usual litany of errors can occur, which must also be reported, and since the NRC estimates that some 60% or more of radiotherapy misadministrations are due to human error, we expect radiotherapy to be simply typical of many such technological systems.

Over the objections and opposition of the American College of Nuclear Physicians and the Society of Nuclear Medicine, the NRC has imposed and actively enforces a Quality Management (QM) rule. The objective was to reduce the number of errors and require adherence to written and auditable QM procedures. As of 1997 the NRC reported that some 138 misadministrations had occurred since the imposition of the Rule in 1992, and that "the net number of reported therapy misadministrations remains at approximately 30 to 40 per year" (SECY-97-037, "Assessment of the Quality Management Program and Misadministration Rule, U.S. NRC," February 12, 1997, p. 11).

This observation, we would argue, is once again a reflection of attaining the minimum error rate, as we have often seen. The Rule by itself cannot remove the errors.

Moreover, the NRC maintains an extremely valuable database of the misadministration incidents, some of which involve multiple patients, and which clearly shows the same litany of errors we have seen before. Thus, we find errors such as miscalculating the treatment size and/or duration; utilizing the wrong source; treating the wrong place; misreading or not following procedures; or even in some recent cases overriding safety interlocks and/or software. These are not necessarily and usually fatal accidents, but simply unfortunate events that happen. These are all errors occurring within the over-all medical system.

We can test the estimates given above for the overall medical error rate against this completely independent set of error data from the NRC. The radiation dose is calculated, controlled, and easily measured, but the occasional errors in the procedure provide reports that should be amenable to analysis.

The error data reportable under the licenses for the interval from 1980 to 1984 were in a report from the event analysis group at NRC (AEOD). There were 310,649 patients treated at some 17 hospitals. Of those patients, there were 68 misadministrations of dose, which is an error rate of ~1 in 4500, or 218 errors per 1,000,000 patients.

We plot this single point on the same graph as all the death data from the studies used by the IOM and shown in Figure 4.5, where the accumulated experience (sample size) taken is the total number of patients treated. Looking at the new graph shown in Figure 4.5, for the accumulated radiotherapy experience (sample size) of about 300,000 patients we might have expected a slightly higher error rate, whereas the point actually falls below the CR line. There are two good reasons that it may be a little lower. First, not all the radiotherapy errors lead to death, as in the IOM data. Second, the radiation dosage is easily determined, and errors could be expected to occur less often for the same accumulated experience.

*We may conclude that the radiotherapy error rate is lower than but consistent with the medical error data, and errors are comparable throughout the technological system, once the effect of the differing accumulated experience is accounted for.*

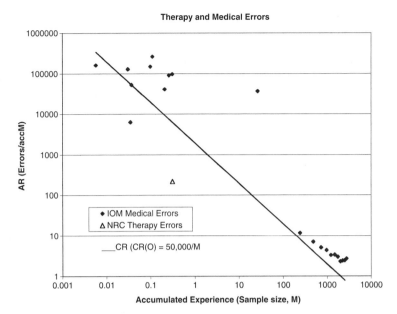

*Figure 4.5 The error rate from radiotherapy (single open triangle point) compared to the IOM medical error data (many filled-in lozenges) from Figure 4.2.*

## Learning Trends Establish a Baseline Risk: Infant Mortality from Cardiac Surgery

There is always a risk of death during or from treatment or surgery. After all, the choice is often between the risk from surgery or prolonged illness or possible death anyway, so we personally and as a society try to balance the risk versus the benefit. This dilemma is true for all medical treatments and becomes very pronounced for surgeries that may have high risk of death. So there are both personal and professional choices and decisions on whether and how to proceed. For the patient, this requires informed consent, based on being told what the risks are of an "adverse outcome."

To determine the risk of death during or after surgery and to aid informed decision making, we need a baseline to establish the normal or usual rate of death from such complex procedures. We believe the DSM provides a useful means to determine that baseline risk.

The situation can be particularly grave for tiny infants born with heart defects, who often may need highly skilled corrective surgical procedures to have any chance of a reasonable life. We were fortunate enough to be made aware of the data collected on infant mortality during the extensive Bristol Inquiry on the deaths of very young (<1 year old) infants.

The Report of the Inquiry (see http://www.bristol-inquiry.org.uk) runs to over 500 pages, makes about 200 recommendations, and found multiple overlapping causes of problems with infant patient care. The Report contains many statements about the importance of learning from errors and mistakes and the need for a culture in the workplace that values learning. These are remarks with which we agree and are indeed fundamentally at the heart of the DSM approach. The Report variously noted and found that:

> Around 5% of the 8.5 million patients admitted to hospitals in England and Wales each year experiences an adverse event, which may be preventable with the exercise of ordinary standards of care. How many of these events lead to death is not known but it may be as high as 25,000 people a year.

> The Experts to the Inquiry advised that Bristol had a significantly higher mortality rate for open-heart surgery on children under 1 than that of other centres in England. Between 1988 and 1994 the mortality rate at Bristol was roughly double that elsewhere in five out of seven years. This mortality rate failed to follow the overall downward trend over time which can be seen in other centres.

> Our Experts' statistical analysis also enabled them to find that a substantial and statistically significant number of excess deaths, between 30 and 35, occurred in children under 1 . . . in Bristol between 1991 and 1995. As our Experts make clear, "excess deaths" is a statistical term which refers to the number of deaths observed over and above the number which would be expected if the Unit had been "typical" of other . . . units in England. The mortality rate over the period 1991–1995 was probably double the rate in England at the time for children under 1. . . .

> Indeed, (one witness) spoke in terms of the "inevitability" of a "learning curve," by which it was meant that results could be expected to be poor initially, but would improve over time with experience. They could argue that the small numbers of children who were treated meant that their figures looked worse when expressed in percentage terms, that they treated children who were more sick (albeit that there was no evidence to support this assertion) and that, once the hoped-for new surgeon was appointed, the pace of improvement would quicken. (Source: The Report of the Public Inquiry into Children's Heart Surgery at the Bristol Royal Infirmary, 1984–1995, "Learning from Bristol," 2001, Chair, Ian Kennedy.)

These extracts confirm that to estimate "excess deaths" we must determine the "normal" or expected typical rate of deaths (the baseline risk). We might expect that to vary by type of surgical procedure. We can now do this using the DSM approach.

The Report contains an Annex B, which includes other papers with a compilation of data from some twelve medical centres conducting cardiac surgery in the United Kingdom. The data files are from two main data sources: the Cardiac Surgical Register (CSR) for 1984–1995, and the Hospital Episode Statistics (HES) for 1991–1995. Some of these under-1 data were apparently analyzed with a best-fit line to provide the baseline to estimate the "excess mortality" at the Bristol hospital.

As always, we are interested in determining if the trend in the data fits with the minimum error rate theory or not, and what the uncertainties and bounds may be in the projections and estimates.

So we tabulated and plotted *all* the data as two sets from the sources in Annex B, for the twelve surgical centers and all procedures, which set covered in two lists in Paper INQ 0045 0057. In the original report the first data set in Figure 6.1 of the report was for open-heart surgery mortality for under 1 year olds only, covering 1350 patients from 1984 to 1995. The second data set in Table 3.3 of the report was for mortalities in the combined data covering 10,361 patents for all open- and closed-heart surgical procedures.

For the appropriate working measure of accumulated experience we used the "patient volume" as given by the Bristol Inquiry Annex, namely, the total number of cardiac surgeries given for each center for each data set for each year.

We calculated the mortality rate as a death rate, D, per total patients, P, or surgeries performed, D/P, and plotted the entire cardiac data sets. We then fitted the sets with the Minimum Error Rate Equation (MERE) as being the baseline risk, which is of the Universal Learning Curve form:

$$A = A_m + A_0 \exp(-P/K)$$

where $A_0$ is the initial rate, $A_m$ the minimum observed or attainable rate, and K the learning rate constant.

The values of the parameters are derived from TableCurve 2D and are given by:

$$\text{Deaths per patient, D/P (all)} = 0.07 + 0.28 \exp(-P/156)$$

This equation implies an initial percent mortality of 28%, which is FOUR times higher than the attainable minimum of 7%.

The results of our data analysis are shown in Figure 4.6.

There is obviously some scatter, but the learning trend is clear. Now the dotted line in Figure 4.6 is the line for infants <1 year old only and is given by:

$$\text{D/P (under-1)} = 0.125 + 0.22 \exp(-P/132)$$

This equation implies an initial rate of 22%, which is similar (after all they both use nearly the same data here), but a minimum rate of 12.5%, which is only 50% of the initial rate.

Therefore, we may also imply that the baseline risk for <1 year olds is generally higher. In both cases the learning rate is similar, that is, an "e-folding" decrease in death rate with a volume of ~140 patients.

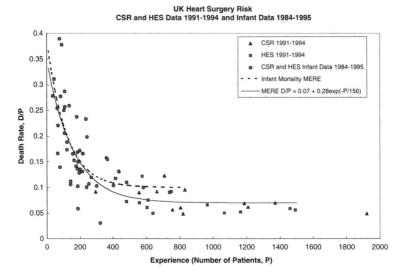

*Figure 4.6 U.K. heart surgery risk: CSR and HES data, 1991–1994, and infant data, 1984–1995.*

The curves at 95% confidence level encompass a band of about ±10%, which is sufficient to encompass most of the data points shown.

We would define or state excess deaths from the DSM as being those rates that lie systematically outside the 95% confidence prediction bounds of the baseline MERE.

We can conclude from these data that there is clear evidence of learning in this highly skilled, specialized, and difficult surgical arena. The statements in the Bristol Inquiry Report that support shared and continuous learning and that "without knowing there can be no learning" is exactly in accord with the precepts and findings of our analysis.

This result of the data analysis from the twelve cardiac centers for 1984–1995 is entirely consistent with learning from our mistakes and errors, and is also consistent with the random occurrence of errors (in this case patient deaths).

The data for the mortality rate of patients follow a learning curve. This curve is consistent with minimum error rate theory that describes accident and injury data from the available human record of the last 200 years.

Due to the statistics of learning from random events, the ratio of the initial and minimum attainable mortality rate for 1984–1995 is about four to one for all cases and two to one for the <1 year olds. The learning rates are, however, similar, and the minimum attainable rates are 7% and 12%, respectively.

As always, additional data would be helpful to reduce the uncertainty bounds and to refine the base equation. But the key finding is that we may use the DSM as a general approach to help to define the baseline risk.

### Disease and Death Declines: McKeown's Hypothesis

We also die from infections and diseases, and one major killer throughout human existence has been pulmonary infections such as bronchitis, pneumonia, and other lung problems. We came across the death rates for 1840–1970, given by Thomas McKeown in his seminal article "Determinants of Health in Human Nature," published in 1978. He argued and showed that "modern medicine is not nearly as effective as most people believe." The major improvements in health, and hence reduction in our risk, came from the major advances in hygiene, nutrition, and safer food and water that have occurred over the past 200 years.

To back up this argument, McKeown used data. We like that approach! It is the basis for our work. The data he gave for various large-scale killing diseases indeed show a dramatic decline over the last century. In fact, the rate was falling very fast, even before the major introduction of vaccinations for smallpox from 1870 onward and antibiotics from around 1930 (when the population was about 2 billion).

In Figure 4.7, we show the "mortality rate" data for pulmonary diseases as the deaths per million plotted versus the accumulated experience—which in this case is the world population. The data are shown as IR values of deaths per million (D/M). The dramatic decline is evident immediately, and the error rate quickly falls below the CR(0) initial value of 3500 deaths per million, but has not approached the lowest AR values. So there is clear evidence of learning, but the rate, dramatic though it is, has apparently not declined as fast as it might have, based on the accumulated experience.

To give some human facts and perspective behind this observation, just over 90,000 people died from pneumonia and influenza in the United States alone in 1998, which is 4% of some 2,300,000 total deaths with a total death rate of about 4700 per million (CDC National Center for Health Statistics, 2001, http://www.cdc.gov/nchs/fastats/deaths.htm). Figure 4.8 shows the IR fit from TableCurve 2D from the exponential MERE model.

$$\text{Deaths per million, IR (D/M)} = 236 + 26740 \exp(-B/0.48)$$

where B is the world population in billions.

The predicted rate today, and for above about 4 billion population, is a minimum rate of ~240 deaths per million. There is 95% confidence uncertainty band that it is between 0 and 500. This minimum rate is about 100 times (two orders of magnitude) less than it was in 1840, reminding us of the large decrease (steep learning curve) in airline crash rates with increasing accumulated experience, and is very close to the radiotherapy

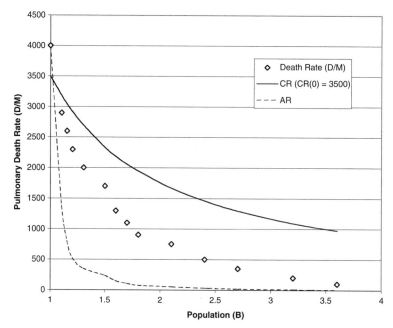

*Figure 4.7  McKeown's data for pulmonary IR death rates, including the AR and CR lines.*

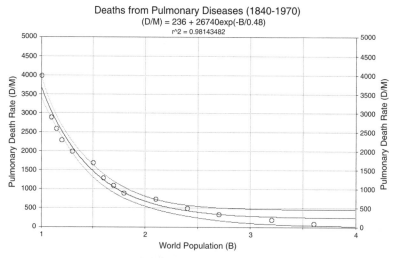

*Figure 4.8  World death rate per million from pulmonary diseases (1840–1970) showing evidence of a learning curve and the possible emergence of an asymptotic rate. (Data source: "Determinants of Health in Human Nature," by T. McKeown, April 1978, as also quoted in L. Horwitz and L. Ferleger, Statistics for Social Change, 1980).*

error rate. *The implication is that with our modern technology we are perhaps now reaching the limits of the reduction in error rate that we may achieve.*

This classic "learning curve" shows that we have learned from the problems and causes of disease and taken the right corrective management actions and health precautions. McKeown called this process "a modification of behaviour" that led to a permanent improvement. We agree, except that we call it simply "learning," and thus error reduction. In this pulmonary disease study, the errors were social errors of poor sanitation, inadequate diet, and lack of preventative health care. As a result of learning and error reduction, the rate of death due to chest illnesses is now very much lower. We now seem to be at the minimum or bottom of the "bathtub" curve where, having removed the predominantly social issues, the remaining errors are probably mainly medical errors, if there is no genetic predisposition.

*So practicing "healthy living" (preventative medicine) reduces the risks of having to subject oneself to medical treatment (engineering medicine).* The biggest improvement is due to social programs that improve sanitation and living conditions. But has the quality of our social environment also improved? Are we also still at risk from other unexpected causes?

## 4.3  TERROR IN TECHNOLOGY

### *Hijackings and Terrorism*

In our modern technological society, there are many other cases where we do not control our own risk and destiny. In the sad and terrible events of September 11, 2001, terrorists hijacked four commercial aircraft, before plunging two of them into the World Trade Center in New York and another into the Pentagon. These were American and United Airlines Boeing 767s and 757s, large and modern aircraft.

These awful events are an extreme example of irrational and entirely incomprehensible behavior, simply designed to disrupt, destroy, and endanger. Airliner hijackings by both terrorist organizations and desperate individuals began long ago, in the early days of commercial aviation, with flight diversions and occasional aircraft destruction and loss of life. It has happened worldwide, to many airlines in many places.

To reduce the threat and risk, technological measures to increase physical security were introduced. These included baggage scanning, subjecting passengers to metal detectors and physical searches, limitations on items that could be carried on board, positive identity checks, and screening, plus higher standards of overall airport security. These technologies can fail and hence are subject to error, or some *confluence of circumstances* that allows access for even known extremists to airports and to pass security devices where humans are deployed, as the events clearly show.

But are you safe? How safe can you be? Can the DSM approach provide any insight here? We need to examine the trends, and once again we turn to the available data.

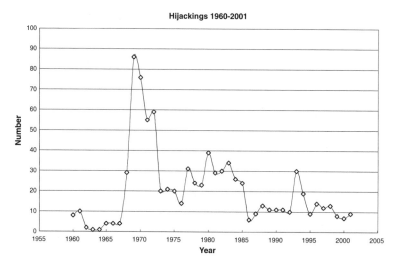

*Figure 4.9  Historical numbers of world airliner hijackings, 1960–2001.*

The Aviation Safety Network contains data for hijackings from 1960, listed by year, and for some 860 events. They are given year by year, and we extracted all these reported events, excluding the unreliable and incomplete data for the USSR and CIS (Commonwealth of Independent States).

We need to take a measure of the accumulated experience, so we take the total number of flights accumulated as an indicator of the learning and error opportunity. The Boeing Aircraft Company gives the worldwide total number of commercial flights for 1965 to 1999. The raw data (numbers of hijackings each year) are shown in Figure 4.9, where we can see that although there are fluctuations and dips in the numbers, clearly hijackings have continued.

The number of flights in a given year has increased by a factor of 10 (from ~1.5 million to over 15 million each year), so to correct for that large experience change the DSM used the accumulated experience in millions of flights (MF), with the result shown in Figure 4.10.

Between 1965 and 1999 there were nearly 400 million flights, and we see that the hijack rate has indeed followed a learning curve worldwide. There is a clear trend with a MERE given by

$$IR \text{ (Hijacks per million flights)} = 0.8 + 10 \exp(-accMF/64)$$

We have shown a projection out to another 100 MF, and it can be seen that the minimum rate is ~1 per million flights, somewhat less than the likelihood of a fatal accident.

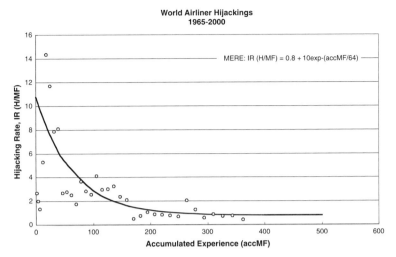

**World Airliner Hijackings**
**1965-2000**

MERE: IR (H/MF) = 0.8 + 10exp-(accMF/64)

*Figure 4.10 The world hijacking rate, IR, shown using the accumulated experience. (Data sources: Aviation Safety Network, 2001; Boeing, 2001.)*

*Therefore, the involuntary risk is negligible.* Nevertheless, many people shunned airline travel in the wake of the 2001 hijackings. Is this also a case of erroneous risk perception at work? Should the risk for security error really be compared to the errors in alternate transport modes, or to one's risk of "a natural" death by other causes? Let us see what the data indicate.

### Being Murdered

Can another totally unexpected social risk happen to us?

As we stated at the very beginning of the book, none of us expect to be murdered; we expect to die "naturally." Statistics show that most murders are committed by people who know their victim and that being randomly killed is less likely. Canada is a good example of a modern technological society, with open communication, a working democracy, a robust legal system, and a stable society with a reasonable standard of living. There is not an excessively large use of the guns and drugs that might be thought to lead to murders, for there is licensed hunting and socialized medicine.

Yet Canada is not crime-free. We wondered how many "societal errors" there were— that is, the risk of being murdered—and whether society is learning to reduce the death rate. We would assume that, for the purposes of our DSM analysis, homicide is itself an error (an accident) and that homicide usually is not meant or premeditated, although obviously sometimes it is. There are also the pleas of "justifiable homicide" or "manslaughter." For the premeditated homicides, we might assume that they are the result of societal error in some way. These rare events could also give us a measure to which we can compare our other error rates.

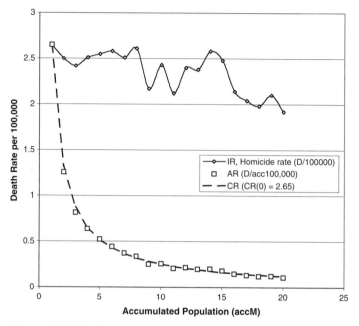

*Figure 4.11  The death rate from homicides in Canada from 1978 to 1997 plotted against the accumulated population, showing the IR, AR, and CR. (Data source: CDC National Center for Health Statistics, 2001.)*

Such data were recently published as a "Homicide Survey" by the Canadian Centre for Justice Statistics for 1978 to 1997. Homicides are called rare because they are about 0.02% of the some 3 million criminal incidents. The rates were given as deaths per 100,000 populations on a year-by-year basis, and we assume for convenience and with lack of an alternative that the risk of homicide is randomly (evenly) distributed around the country. We need a measure of the accumulated experience, so we take the total accumulated population as the basis.

Canada has about 30 million people. The population data are available as official census information from Statistics Canada, so we used those data and linearly interpolated (filling in the missing years) between censuses. The result of the DSM analysis is shown in Figure 4.11, where we have plotted as usual the IR, AR, and CR, the last being based on the rate in 1978 of ~2.65 per 100,000.

To a first approximation, the rate is constant and tracks the CR closely. The implication is that there is little learning and that the rate is close to a minimum. The IR wanders around, somewhere between 2 and 2.5, with a slight downward trend, so it is difficult to make an accurate prediction. The AR curve follows the CR so closely that it implies that the rate is likely to be the minimum for this technological society. Either the minimum has indeed been reached or the societal systems are not producing effective further learning (homicide reduction).

Since the minimum rate is ~2 per 100,000, we can convert that to a "homicide error interval." If we take any assumed group of 100,000 people and the 100,000 humans all live for 1 hour, we have 100,000 hours of experience. *The chance (probability or risk) of one homicide is then ~2 in 100,000 hours, or an interval of 1 every 50,000 hours.*

That's about two to four times higher than our chance of being killed by error flying with a commercial airline. We should not fear flying on a relative or rational basis.

## 4.4 DRINKING WATER: WALKERTON AND SYDNEY'S RISK

We now turn from acts of terror and murder to nonviolent deaths, due to errors in our social and governmental regulated systems.

How safe is your water?

We all expect to have safe water to drink, although actually for a large percentage of the world's population that is a luxury. But in modern technological systems, we know how to treat, pump, and deliver safe drinking water; it is considered a natural thing to have confidence in the water. As we were writing this book, the Concorde crash occurred—and so did the tragedy of Walkerton. In this rural community in Canada, the water became unsafe to drink through the presence of *E. coli* due to ingress of contamination into the wells and the lack of a working chlorination system.

Seven people died, thousands were sick, and the outcry was immense. Compensation to all the 5000 or so affected residents of the town was offered at a level of several thousand dollars each. The total cost was estimated at over $60 million Cdn. Many residents still will not drink the water, not believing it safe.

A Commission of Inquiry was rapidly established, a procedure that often follows such headline-making events. The terms of reference were simple:

The commission shall inquire into the following matters:

(a) the circumstances that caused hundreds of people in the Walkerton area to become ill, and several of them to die in May and June 2000, at or around the same time as *Escherichia coli* bacteria were found to be present in the town's water supply;

(b) the cause of these events including the effect, if any, of government policies, procedures and practices; and

(c) any other relevant matters that the commission considers necessary to ensure the safety of Ontario's drinking water, in order to make such findings and recommendations as the commission considers advisable to ensure the safety of the water supply system in Ontario. (Source: http://www.newswire.ca/releases/June2000).

## *The Search for the Guilty and for Closure*

Everyone in Canada wanted to know why and how this water contamination could have happened; the townspeople needed to determine guilt and to obtain what is now called "closure" for all those who had seen relatives, friends, or family suffer. We would refer to this as also learning from the errors.

As we wrote, the Public Inquiry into the Walkerton water tragedy was completed, having gathered the required reams of information. The Inquiry is really intended to be sure that this "cannot happen again," this being the earnest and thorough search and quest for "zero defects." *But testimony given to date shows the usual confluence of factors in the operation of the water treatment plant and in the testing and reporting of water samples.* The staff managing the water treatment, who were also very upset by what had happened, said basically that they were just unaware of the dangers, had not been trained to recognize the importance of *E. coli* contamination, were confident in the water quality, and made a series of poor decisions.

The quotes from the transcript of the questioning of the General Manager of the water treatment plant by an inquiry lawyer are illuminating in showing up the subtle error factors. They are in Appendix B.

The Ministers and the Premier of Ontario were also called to testify, since as a political action the "privatization" of drinking water testing had occurred under the government and budgets had been cut. The question in many people's minds was whether this had led to errors, since accreditation and mandatory reporting had not then been required (compare the case of U.K. Railtrack).

In one response the Premier said:

> A: . . . let me repeat that had I or both . . . Ministers that you heard from, or senior staff believed that there's any risk in this plan to human health or any increased risk, I—I—I think I've got to repeat that

> Q: Right.

> A: —there's risk in everything, there's risk in—walking across the street, that any of these actions would increase risk, had we believed that, we wouldn't have proceeded.

And then:

> And as I've indicated to you, there was an attitude, certainly, that had been generated over a period of time that more money meant better protection for the environment. We rejected that. (Walkerton Inquiry Transcript, June 29th, 2001, Examination-in-Chief of the Premier)

The facts are clear: adequate chlorination kills *E. coli*, and obviously the water was undertreated. Chlorination plants and water treatment are a technological system that is well understood and not too complex, but does require human interaction, training, and knowledge. Thus there were multiple context, organizational, and environmental contributing causes, and a *confluence of factors* such as:

1. Distraction and time pressures
2. Poor training
3. Violation of procedures
4. Poor decision making
5. Unverified test data
6. Unawareness of danger
7. Inadequate management oversight
8. Lax regulation and reporting
9. Lack of quality assurance
10. Overall working environment
11. Lack of automatic warnings

These contributory factors and warning signs for technological systems with human interaction are by now all too familiar to us: every event and error we have studied contains the same litany of causative factors. Perhaps they are in different words or different order, perhaps in different combination, and perhaps of different importance.

You might think that we would expect such errors to occur by now, and be sure that the appropriate systems are in place to provide error reduction through learning. *We should learn from our errors, measure our errors, and ensure that we are learning and reducing errors.* What is done in practice is often piecemeal. Investigation is often thorough and in great detail and expense. But inquiries are conducted case by case, error by error, jurisdiction by jurisdiction, agency by agency, without realizing the systemic and underlying causes and their cross-cutting relationship to other errors. *So time and time again, we are apparently doomed to rediscover the same elements of error causation, hence ignoring their wider ramifications.*

### Australia's Water

But extensive water contamination had happened before, still happens, and is well documented. Water contamination occurs not just in small rural towns such as Walkerton, but also in large modern cities with all their technological infrastructure. In Australia, evidence of contamination of the City of Sydney's water supply by the organisms *Cryptosporidium* and *Giardia* was evident in late July 1998. The Health Department was informed of the event but the levels did not raise health concerns, since low levels of such organisms are commonly found in water supplies throughout the world. By Sunday, July 26, high readings (some extremely high) were found, and by the afternoon of Monday a boil-water alert, a common response, was issued. It was then believed that the contamination was perhaps due to a combination of a broken sewer main and a broken water main, perhaps from major construction projects. Low levels of contamination were obtained on Tuesday and Wednesday, but by late in the afternoon and early evening, high readings appeared at a filtration plant and a reservoir and some locations further down the system. By Thursday, high readings were obtained from water sampled elsewhere and a Sydney-wide alert was declared.

Public concern was raised so high it required a ministerial response to the problem. The water supply was declared safe on Tuesday, August 4. The government formed an inquiry on August 5 to investigate the possible cause of the contamination and the public communication (emergency notification) procedures used. Following this first major event, which occurred between July 21 and August 4, 1998, on Thursday, August 13, higher levels were measured downstream of a treatment plant. Embarrassingly, there was a second event from August 24, 1998, and a third commenced on September 5, 1998. Some of the readings were so high as to be doubted, nationally and internationally.

Although it was not thought that anyone died as a result, the inquiry was extensive and thorough, since it was regarded as unacceptable that a component meant for the treatment or distribution system might contribute contamination to Sydney's water supply. After all this work, and five reports, the inquiry considered all the possible causes suggested. None could be ruled out, but some were considered unlikely. Others to "varying degrees were possible *and may have operated in combination to cause the event*" (emphasis added). So we have a *confluence of factors* and latent (undiscovered) errors.

The final report is very clear in defining the state of ignorance:

> The present level of scientific knowledge makes it impossible to identify all the factors that have contributed to the contamination events and to meaningfully predict the likelihood of its recurrence. However, after the experience of the recent events and the implementation of the various proposed management changes, [the Inquiry] is confident that Sydney's water supply can be managed to minimise all reasonable risks to public health. (5th and Final Report of the Inquiry, http://www.premiers.nsw.gov.au/pubs.htm)

So once again we find that we cannot eliminate the errors and the problem. But can the risks indeed be minimized? Can we manage errors? Will the errors, whatever they were, happen again? And if not in this water system, perhaps somewhere else?

What do the data tell us? *How safe are you?*

### Drinking Water Incidents: Errors in Water Treatment

We looked around for good and extensive data on drinking water problems, for the types of systems that are essential in supporting a modern technological society. We were made aware of the exhaustive work of G.F. Craun, who reported the extensive U.S. water "event" data from 1920 to 1997.

In the Craun data sets are water contamination events that caused illness outbreaks for *E. coli* and *Cryptosporidium*, such as we have just discussed, and many more types such as typhoid, gastroenteritis, and hepatitis. The data reported by Craun were grouped in intervals of 10 years, for treated and untreated water, for groundwater, and for distributed (centrally managed) systems. Treatment here refers to the active application of filtration, purification, and chlorination technology.

Over the 77 years, a total of some 1900 events (outbreaks) were reported. The outbreaks caused nearly a million illnesses (~890,000), which is an average rate of more than 20 events a year affecting more than 450 people each time. In that sense, the Walkerton incident is unfortunately an extremely typical, but not extraordinary, error in North America.

The causes of all the U.S. events were not clear, although the consequences of illness, death, and boil-water advisories were. The immediate thought is that event rates for treated water should be lower than for untreated water, so we calculated and plotted all the data using the DSM. Since the total number of water sources was not given, the choice for the accumulated experience was based on the *total population*, on the reasonable assumption that practically everyone drinks water. The rate (IR) was similarly calculated, as the choice of personal water source would be dictated by where a person happened to live, with the chance of an error in treatment (an event) affecting that individual being nominally the same for everyone then living. The population grew by more than a factor of 2 over the time frame 1920 to 1997, reaching an accumulated population of nearly 1.5 billion on which water treatment and consumption had been practiced.

Contrary to our initial expectation, we see immediately in Figure 4.12 that the rates fluctuate around *and do not depend much on whether there is nominally water treatment or not*. There is some evidence of a decreasing rate or slight learning curve for all sources. We now know why there is no significant difference: the outbreak events are due to errors that are independent of the *initial* presence of treatment, as exemplified by both the Walkerton and Sydney "events." Despite the presence of modern technology, errors can still be made, water sources can be contaminated, and perfect treatment is not apparently possible.

In fact, events occur all the time, due largely to human errors that lead to contamination of either the source or the supply. The event (and hence error) rate is not declining as fast as the constant rate expectation either, a simple fact that is demonstrated by comparing the AR to the CR. With an event rate of ~1 in 1 million of the population, this is a personal or societal risk level that is comparable to that of being struck by lightning. But we cannot determine the actual error interval from these data, since we do not know how many wells, sources, facilities, staff, and treatment operations are involved.

We can make a "guesstimate" if we boldly *assume* that there are about 4500 people drinking from each water system (i.e., 10 times the average number who get ill for each outbreak). You may even want to choose your own estimate. This assumption gives

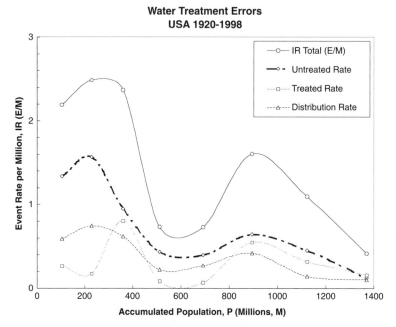

*Figure 4.12  The rate of water contamination events in the United States in the 20th century. (Data source: Craun, 1991.)*

~3300 water systems for an average total water-drinking population of, say, 150 million. So we have accumulated

$$\sim (3{,}300 \text{ systems}) \times (365 \times 24) \times (1997 - 1920)$$
$$\sim 2 \text{ billion water system-hours of operation}$$

If there is one person responsible for treating each such system, then the average error rate per outbreak for the 2000 reported events over the interval 1920–1997 is very roughly

$$2000/2{,}000{,}000{,}000 \sim 1 \text{ in 1 million hours of operation}$$

Since the assumed one person works 8 hours a day, or about one-third of the time, the error rate is given by

$$\sim 1{,}000{,}000 \text{ operating hours}/3 \sim 1 \text{ in 300,000 hours}$$

Based on the DSM analysis of active errors, which gives ~1 in 200,000 hours, we would surmise that this interval is perhaps low. There could be many errors in this estimate, too, but we do not even have to claim high accuracy for this type of calculation.

It is intended only to provide a rough answer and to show how the calculation might proceed if we had the type of data that is needed.

We could postulate that this error estimate is the right order of magnitude for contributions from "passive" errors, where something should have been done and was not. This is an error of omission. In this case, the passive error is forgetting to treat the water, or allowing contamination to enter the source, or adopting inadequate filtration, monitoring, or purification methods for detecting degrading water source conditions. The technological system is meant and expected to just keep functioning, and we become complacent and forget to tend to the machinery and its source material correctly. We might expect passive errors to be less frequent than active errors, since we more usually have an active interaction with the technology, in its operation, maintenance, design, and manufacture.

A passive error is inherently different from an active error, as the latter requires something (a task, an action, a procedure) to be not carried out correctly. In the terminology of human factors research, passive types are called "errors of omission," including forgetting to do something, versus active types, which are termed "errors of commission" and are made while actually doing something.

### Dam Failures: Passive Errors

In Appendix A, we elaborate the MERE theory to include forgetting as a natural extension of the analysis of Universal Learning Curves. The theoretical result shows that the *ratio of forgetting to learning is the key*. Hence the observed overall learning curve actually may depend on how fast we are passively forgetting versus what we should be actively learning (i.e., the ratio of the passive to active error rate constants). But are there other possible examples of passive errors? *Are you safe* from such errors?

Our drinking water may not come from a well; it may be from a reservoir, where the water body is formed by a man-made dam. Although floods are called natural hazards, dams are a technological system and are an invention where humans bear the total responsibility for changes to the natural water flow. A dam is a barrier built across a waterway to control the flow or raise the level of the water. Dams can fail catastrophically, as they have over the past 200 years. Dams give us the opportunity to examine a technological system that is widely deployed, but where human interaction over the life of the dam is very small, and the failure process is largely due to the absence of the needed human intervention.

The worst disaster caused by a dam failure in the United States occurred in the spring of 1889 in Johnstown, Pennsylvania. Following a night of heavy rain, the South Fork Dam failed, sending 20 million tons of water and debris down a narrow valley onto Johnstown. The flood caused more than 2200 deaths and left many more homeless. "We didn't see it but we heard the noise of it coming; it was like a hurricane through a wooded country; it was a roar and a crash and a smash," said John Hess, a survivor.

Dam failures still occur today:

> Yesterday the state of Mississippi reported a dam failure in the community of Byram just south of Jackson in Hinds County. The levee started to break due to heavy rains, with the break located in the center of the dam. Approximately 50 families were evacuated for the day from two subdivisions across from the lake. (Federal Emergency Management Agency [FEMA], January 23, 2001)

Dams fail for a variety of design flaws and other factors, often acting in confluence. In the United States, the Federal Guidelines for Dam Safety, which are followed by all federal agencies responsible for design, define the construction, operation, and regulation of dams. They are defined as being over 25 feet in height and holding more than 50 acre-feet of water. "As existing dams age, dam safety has become an extremely important issue," says the United States Committee on Large Dams.

Many U.S. Federal Agencies own or operate dams, or have an interest in dam safety. The 1996 update to the U.S. National Inventory of Dams (NID) lists 75,187 dams, and in 1999, FEMA listed 76,750. Of this total, some 80% are earthen dams, and more than 95% are owned by the states, local governments, industry, and individuals. Of that number, about 10% are used for providing drinking water, and more than 10%, or about 9000, are listed as "high hazard," meaning human life would be endangered by dam failure.

Floods are a known hazard, and flood insurance is a recognized requirement for property owners. Away from the coasts and flood plains, the flood hazards that pose the greatest risk to human safety are due to tsunamis and dam failures, and these hazards are not shown on current Flood Insurance Rate Maps. A recent issue paper written by the U.S. Bureau of Reclamation for the Western Governors' Association

> spells out in alarming detail the hazard posed by the normal operation of some dams. People have built homes and businesses downstream of dams that for years have reduced the peak discharges to the streams or rivers upon which they are constructed. The new development has severely restricted the ability of the dam operators to pass flood flows, causing the dams to now pose a greater threat of potential failure. (FEMA, 2000)

It is possible to expand the data in the original paper Maps using modern digital technology as a database tool to include other hazard information. In the meantime, we are learning from technological system failures; extensive guidelines on dam design and construction, monitoring, and maintenance, and on hazard evaluation, are available from FEMA. A database now exists on dams in the U.S. National Dam Performance Program (NPDP) that was set up in 1994, and there is a National Dam Safety Program under FEMA.

In Europe, there was a great outcry and concern over the failure in early 2000 of two earthen dams in Romania (at Baia Mare and Baia Borsa). Containing mine tailings, these dams were designed to stop any materials flowing into the Danube River basin. The dam failures were due to heavy rains and snows and released hundreds of metric tons of cyanide and other heavy metals, seriously damaging marine life. The accidents were stated as caused by inappropriate design, regulatory approval of that design, and inadequate monitoring of the dam construction, operation, and maintenance. These reasons sound very familiar by now.

But what do the data tell us? Are dam failures decreasing or not? Are we safe?

Some years ago, Karl Ott, who is an unsung pioneer of statistical error analysis, examined dam failure data from 1850 to 1980. The seminal report from 1984, made available to us by Ott, was published in Germany and co-authored with Hoffman and Oedekoven. The analysis showed a steadily decreasing number of failures per dam-year with increasing operating time. That trend is exactly what we expect from the DSM learning-curve approach, so we eagerly reexamined the latest available data.

The Department of the Interior's Bureau of Reclamation has an active Dam Safety Office that has an excellent research activity. Tatolovich's comprehensive 1998 Bureau report gives dam failure probabilities for specific dams that range from 1 in a 100 to 1 in 100,000 years for dams that are up to some 100 years old. Appendix J1 of Tatolovich's report lists the so-called von Thun failure probabilities. These failure rates are based on the historic dam failure rate data grouped by type of dam and giving the life-years, which are exactly the IR and the accumulated experience in dam-years, respectively.

We plot these data in Figure 4.13, covering all dam types and failure modes, which gives a maximum accumulated experience to over 100,000 years for the oldest earthen dams.

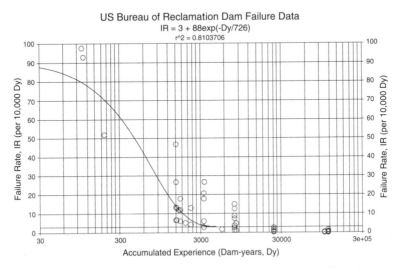

*Figure 4.13 Failure rate for dams in the United States as a function of accumulated experience. (Data source: Tatolovich, 1998, Bureau of Reclamation, Dam Safety Office.)*

The DSM exponential model using TableCurve gives the failure rate as

$$IR \text{ (dam failures per 10,000 years)} = 3 + 88 \exp(-accDy/726)$$

which implies a minimum dam failure rate of ~3 per 10,000 dam-years.

The data for earthen dams for the U.S. Bureau of Reclamation have also been given by Tatolovich, as a "risk rate" based on the failure and incident data. These dams also follow a learning curve, which as we might expect, is a mini-curve within the total population of dams. The MERE for that data subset is given by

$$IR \text{ (earthen dam incident rate per 10,000 years)} = 7 + 110 \exp(-accDy/3632)$$

This implies a minimum risk rate of about twice that for the total dam experience, although the differences are actually well within the data uncertainty.

To be more complete, we also looked at the recent available survey of more than 2000 dam incidents in the United States reported by FEMA for 1989–1998 using the NPDP database. Incidents include failures, incipient failures, and precursors to failure, as an approach to trending, prediction, and flood prevention.

The guidance of the definition for what constitutes a reportable incident is given in Table 4.2. As for the other technologies we have studied, with incidents such as near misses and significant events, there are many more dam incidents than dam failures. To be consistent with the failure data, the basis used for estimating the accumulated experience is the operating dam-years for the entire average population of ~77,000 dams in the United States over the reported interval from 1989 to 1998.

The incident data give very instructive and illuminating results and are shown in the standard DSM format in Figure 4.14. Looking at this figure, it is clear that the incident rate has been increasing, and a learning curve is not being followed. There are apparently 10 times more incidents (1 per 1000 dam-years) than the minimum rate of dam failures (1 per 10,000 dam-years) at this stage of accumulated experience. The upward trend in the incident rate is a wear-out effect: apparently, a minimum rate of ~1.4 per 1000 dam-years has been reached and since exceeded. There is no further decrease yet below the prior minimum, as the tracking of the AR with the CR line clearly shows.

The minimum incident rate is ~1.4 per 1000 dam years, or

$$1.4/(1000 \times 24 \times 365) = 1 \text{ in 6 million hours of dam existence}$$

This incident rate is one of the lowest found for any technological system we have yet studied. The result indicates that the passive nature of dams—with little human intervention after they have been built—provides a possible lower bound to technological system error performance when the *active* interaction of humans with and in the technological system is minimized. We term this the minimum passive error rate.

*Table 4.2  NPDP Guidance for Determining If a Dam Incident Has Occurred*

| Key Words | Incident Category |
|---|---|
| Inspection Findings | The findings of a dam safety inspection that identifies a previously unreported incident of unsatisfactory or unsafe conditions at a dam (exclusive of ordinary maintenance and repair and findings of inadequacies relative to current design criteria). |
| Damage, Signs of Distress, Instability | Observations of damage, signs of distress, or instability of the dam appurtenant structures. |
| Dam Breach, Dam Failure | Any event resulting in the breach of a dam (partial or complete). |
| Controlled Breach | Planned (non-emergency, non-incident initiated) breach of the dam. Possibly carried out to remove the dam from service or to make major repairs. |
| Downstream Release—Controlled or Uncontrolled | Uncontrolled release of the reservoir (e.g., appurtenant structure misoperation), or controlled release with damage. |
| Inflow Floods, Earthquakes | Performance of a dam (satisfactory or unsatisfactory, anticipated or unanticipated) generated by a nearby seismic event or inflow flood. |
| Mis-operation, Operator Error | Mis-operation of appurtenant structures such as failing to comply with the project rule curve. |
| Equipment Failure | Failure of mechanical or electrical equipment to perform the dam safety functions for which they were intended. |
| Deterioration | Deterioration of concrete, steel, or timber structures that jeopardized the structural/functional integrity of the dam or appurtenant structures. |
| Dam Safety Modification | Modifications to improve the safety of the dam or appurtenant structures such as might be required due to changes in the design criteria. Note: Repairs following an incident are reported as part of a follow-up report. |
| Reservoir Incidents | Events that occur in the reservoir (e.g., landslides, waves) that may impact the safety of the dam. |
| Emergency Action Plans | Implementation of the Emergency Action Plan (or emergency actions) in part or whole. |
| Regulatory Action | Regulator has determined an unsafe condition exists, or the dam does not meet applicable design criteria (e.g., inadequate spillway capacity), and requires action to be taken by the owner (e.g., reservoir restriction, safety modification). |

Source: National Performance of Dams Program, 2001, http://npdp.stanford.edu/.

Basically, the DSM analysis indicates that we have indeed learned how to build dams and operate them such as to reach a minimum failure rate. When "natural" aging effects and human inaction become significant, the incident rate rises, which explains why the incident rate in Figure 4.14 shows an upward trend. By careful monitoring and by upgrading of the structure as it ages or weakens, the failure rate can be reduced toward

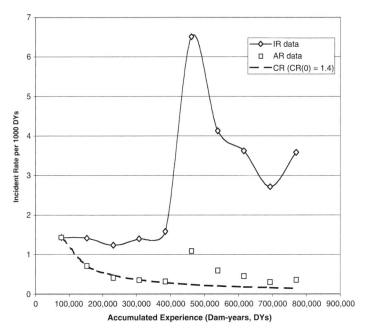

*Figure 4.14 The IR, AR, and CR dam incident rates in the United States. (Data source: FEMA, 2001.)*

the minimum, and the active human error component is in the failure to take such measures. The insufficient learning in this case would apparently be the lack of needed and necessary preventative actions, either by intent or through ignorance.

So, ordinary water is a hazard: we expose ourselves to risk and error by sailing on it, drinking it, or living near a dam. We follow a learning curve with accumulated experience. The lowest rate of "passive" errors exists for structures such as dams where humans are involved mainly in the design and construction, technological systems that we build and then largely leave alone. The lowest rate for "active" errors is about 10 times higher, where humans are also involved in the detailed and everyday operation of the technological system, even just by living in society. So human "hands off" in the everyday operation seems to be better than human "hands on."

## 4.5 ACCIDENTAL DEATHS AND INCIDENTAL INSURANCE

Of course, we could suffer the more usual fate of "accidental death," rather than being murdered. There are many almost silly things (errors) that can happen: falling off a ladder or out of bed; tripping on a step or over the dog; electrocuting yourself while changing a light bulb or mending an appliance; hitting your head on a beam; cutting a finger or a vein; or burning yourself when cooking. There are many accidents in the places where we live and play, and these may cause just bumps and bruises, but can also produce broken bones, infected wounds, bleeding, and even death.

A whole industry is built on avoiding risks and taking safety precautions around the home. Almost all home appliances with any dangerous potential are checked to approved standards, by reputable organizations such as Underwriters Laboratory and government establishments. The errors and accidents usually come from our misuse of the technology: we can sue someone for liability and damages if it was simply unsafe to use when we bought it. The potential product liability is therefore a real issue for any manufacturer, and testing for and warning users about safety are a must.

But fatal accidents still happen, with tools, saws, ladders, mowers, machines, toys, and electrical appliances. Almost any tool or piece of machinery can kill or injure you if not handled properly. They are an integral part of our existence in the modern techno-logical world, where the comparative risk from living can lead to an involuntary death. It is often stated that there are more accidents in the home environment than in any other area of life.

So what do the data say? Are you safe?

We wanted to examine the errors due to accidents in a modern technological society. Recently, as part of reporting on nationwide health care, the data for deaths in hospi-tals in Canada were given as part of the "Health Indicators 2000." The death rates were compiled on a province-by-province basis and covered many causes, such as pneumo-nia and cancer, and also from "unintentional injuries" (that is, for all accidents). The unintentional deaths are defined as the "age-standardized rate of death" from uninten-tional injuries per 100,000 population, where unintentional ("accidental") injuries include injuries due to causes such as motor vehicle collisions, falls, drowning, burns, and poisoning, but not medical misadventures/complications. The intent is that this death rate

> measures long-term success in reducing deaths due to unintentional injuries, compared
> with other regions, provinces, and countries. Measures the adequacy and effectiveness
> of injury prevention efforts, including public education, community and road design,
> prevention, emergency care, and treatment resources. (CIHI, 2001)

In other words, it is a rate of societal learning and error reduction. To use the DSM approach to analyze the hidden trends, we need a measure of the accumulated experi-ence. We chose the province population in millions for the year reported, on the rea-sonable assumption that everyone was exposed to the risk of unintentional injuries in that year. We found the numbers in millions, MP, for the population of each province reported separately by Statistics Canada.

We also had available the industrial injury rates for the years 1993–1997, for the same 13 provinces, from the Human Resource and Development Commission (HRDC).

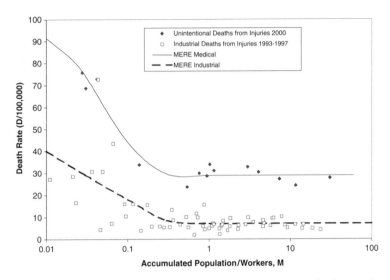

*Figure 4.15 Accidental (diamonds) and industrial (squares) death rates in Canadian provinces, each data point being from a province, and also showing the MERE exponential model equations. (Data sources: CIHI, 2001; HRDC, 2001.)*

To analyze that data using the DSM, we assumed the relevant measure of experience should be based on the accumulated millions of worker-years, MWy, for each province.

The populations varied throughout the 13 provinces in Canada from 30,000 to 11 million, a factor of about 400, so the injury (error) rates also had a large variation because of the very different accumulated experience. We took the "unintentional" and industrial death rates and plotted them versus the accumulated populations, with the result shown in Figure 4.15.

The MERE equation for the accidental death data due to unintentional injuries is given by

$$IR \text{ (Injury deaths/100,000)} = 29 + 73 \exp(MP/0.06)$$

where MP is the millions of population.

The MERE for the deaths due to industrial injuries is

$$IR \text{ (Injury deaths/100,000)} = 7 + 30 \exp(accMWy/0.1)$$

where accMWy is the number of millions of worker years.

As can be seen, the provinces with the lowest experience base or sample size (population) have a rate some three to four times higher than those with a greater accumulated experience (sample size and/or numbers of people). The industrial injury rate is lower than the "unintentional" rate, but the decrease with experience, or the learning constant, is similar for both types of injuries. The ratio between them becomes slightly less with increasing accumulated experience (population).

The total "accidental' minimum error rate reached by the year 2000 in the largest samples is ~30 in 100,000, and ~7 in 100,000 for industrial injuries.

The DSM is a new way of presenting such varied death and injury rates across different sample populations. This experience-based plot of accidental injury rates is similar to the state-by-state analysis and plot of the Australian traffic accidents in Chapter 3: those with the lowest accumulated experience have the highest error rates.

Insurance companies offer accidental death life insurance policies, and they can be very cheap compared to normal insurance, except for certain age groups and employment types. The DSM analysis implies that at least some of the apparent variation in risk—and hence in potential policy premium cost—between subsets of populations (e.g., by location or age) can be partly attributed to differing sample sizes and/or differing accumulated experience.

Thus, we have confirmed, using the DSM, that the risk of being injured and killed by accident in just living in a modern technological society is significantly greater than that of being killed at work.

# 5

# ERROR MANAGEMENT: STRATEGIES FOR REDUCING RISK

*"Accidents are rarely caused by the last control input. They're usually caused by a chain of events."*

—George Jonas, *Southern Newspapers*, 2001

We have now established a much clearer and more consistent basis for error analysis, tracking learning, defining minimum error rates, and intercomparing different technological systems. The approach we have used in the DSM has been to analyze the world's data, determine trends, and compare them to the minimum error rate theory.

What action plan can or should now be carried out to actually reduce risk and increase safety? Are current approaches and techniques right? How safe are you? *Or equally important, since we now know that humans cannot reduce errors and the subsequent risk to zero, how safe can you be?*

## 5.1 QUALITY MANAGEMENT AND THE DREAM OF ZERO DEFECTS

Minimization of errors is an important goal. In the lexicon of the world of "quality" management, the goal of "zero defects" is stated as the output for any process and its improvement. Since defects are nonconformances to requirements, and quality is defined as meeting the customer's requirements without defects, the logical assumption and motivation is and must be the achievement of "zero" defects. Thus, zero defects effectively means no errors and a perfect process.

Thus "zero defects" becomes what is termed the performance standard, what we would call the expectation, and any departure from that is deemed unacceptable, by definition and edict. This de facto "no errors" standard can be used for perfecting production, for process improvement, or for safety management.

In fact, this belief motivates many, and in a recent letter accompanying their safety newsletter a major company states:

> We believe that zero injuries is an attainable goal anywhere in the world and in any industry. It's a belief that has been made a reality at not only DuPont sites but facilities of other companies around the world. (James A. Brock, DuPont Safety and Environmental Management Services, October 1998)

The article goes on to state that *many* sites have "worked literally millions of man hours without a lost workday case." Such rates are not inconsistent with the chemical industry (and other ACSI) minimum injury rates shown in Figure 3.7. With a minimum error rate of order 1 in 200,000 hours, some sites statistically could have an IR lower, and some an IR higher, but not *all* the sites all at the same time.

This statement is a clear example of a company that emphasizes safety and is apparently well down toward what we call the minimum of the learning curve at the stated sites, using "zero defects" (in this case zero lost workdays) as a management goal and with tools and techniques in support of this process.

For those who might doubt the achievement of such a lofty goal, the oft-cited example is the payroll process, where mistakes supposedly cannot be tolerated. Zero defects are therefore a requirement. But as we know, infrequent errors do occur in payrolls, through miscalculating hours, or the wrong deductions, or registering payments to the wrong account. They are quickly corrected, and the payroll process is usually highly automated to try to minimize such errors. The payroll process has a strong and direct feedback from learning.

Recently, in Canada, it was made a requirement to register existing firearms by submitting an application form and a photograph of the applicant. In the startup phase of the computerized system (to December 2000) the firearms license issuance system was found and stated to have a human error rate of ~1%, which in a million applicants meant an error rate of 10,000/M, remarkably and probably not coincidentally close to the rate for medical errors. In both cases, none of us would expect such socially important activities to have such apparently high rates. We are surprised by them.

Now we are certain that errors occur all the time and cannot be reduced to zero. We know this by just looking at the vast data we have studied, covering millions of flights, billions of miles traveled, millions of hours worked, hundreds of millions of people, thousands of industries (some in great detail), billions of operating hours, thousands of deaths, and millions of errors. *All technological systems have errors and allow mistakes: so why do quality managers cling to this concept of zero defects?*

There is clearly comfort to be taken in the quest for perfection and in striving for this idealistic goal.

## *Tracking Performance*

The answer is that in seeking improvement they are really taking advantage of the Universal Learning Curve (ULC). By driving errors down the learning curve using a so-called "quality" or "safety management" system, they are going toward the minimum as rapidly as their particular approach can go. With any new system or process, the initial (startup) apparent error rate will be high, until experience is gained (by learning), and then the error rate (defects or nonconformances) decreases.

By measuring the defects (errors) and tracking them, *this approach is really using common sense.* Unless a good database for the error rate is obtained, you cannot tell if the rate is improving as fast as it might. This is exactly the analog of the IR, CR, and AR analyses we have performed using the DSM. *But we expect the shape of the learning curve (defect reduction rate with increasing experience) to follow the MERE exponential model. We also expect the final error rate to be small, but not zero.*

*Our extensive DSM analyses of error rate data published worldwide show the minimum error rate model to be correct and "zero defects" to be actually a desirable objective but a complete illusion.* There are still many valuable and various tools and techniques that have been developed to pursue the ULC. They track defects and nonconformances rigorously. None of the "quality management" approaches that we know of have the necessary quantitative (data and number–based) tracking versus the "expected" rate.

We have been accused of being pessimistic (as has Perrow, 1984), in the sense that errors are inevitable and are decreased naturally by learning. Why bother, then, to try to reduce errors further or faster, since we are doomed to always have errors?

We are actually optimists! *Errors can always be reduced to the minimum possible consistent with the accumulated experience by using effective error management systems and tracking progress in error reduction down the learning curve.*

By identifying the minimum rate, then quantifying that rate and the learning curve, we can determine quantitatively by rate comparisons the best system for error reduction, and hence *quickly obtain and retain the minimum rate.* What happens with many systems is that, after a period of "good" (low error rate) performance, and usually after a problem or an interval of high attention to the system, they suffer wear-out.

Errors then grow as the system relaxes, and forgetting occurs. This is equivalent to the so-called "bathtub" curve, where failures (errors) increase after a while in a mechanical system as the components wear out and/or maintenance is relaxed. In learning terminology, this is a "forgetting curve."

## 5.2 A Learning Environment: Safety and Quality Management

We suggest using the DSM in a number of specific ways to improve "quality," to reduce errors and defects, and to monitor performance. Just like quality management, it is data-based and -driven. Thus we suggest the following approach to recognizing and managing errors.

Practically speaking, it would seem that an individual organization, industry, or regulator could track the relative trends in accident (error) rates. This would determine at least the following aspects from risk or safety management perspectives as to whether a learning environment exists:

(a) whether the observed error rate is consistent with the average or overall rate
(b) whether significant departures are occurring, and why that might be
(c) where the overall entity's error performance resides on the basis of both the accumulated experience and the overall trends
(d) what the expected present and future error rates (and management goals and expectations) might be
(e) whether the particular technology has attained or demonstrated a minimum achievable rate
(f) whether changes to aspects of the technology, process, manufacturing, management systems, and reporting are desirable
(g) whether the overall error rate is giving acceptable or unacceptable results
(h) whether the adoption of given targets are even achievable without changes in the existing system or technology

This is process management and improvement in action, and it is entirely consistent with a quantified approach to error management, defect reduction, and quality improvement.

## 5.3 Measuring and Managing Our Risk: In a Nutshell

We can now summarize all that we have found out about the minimum error rates, recalling from our text the calculations and estimates from the active error observations. The analysis is given in Table 5.1, where the error frequency is compared for a number of different risks.

These results use data that cover 200 million flights, hundreds of millions of patients, about 1 million ship-years, more than 20 million workers, thousands of millions of vehicle miles, thousands of millions of train miles, millions of components, and thousands of homicides and accidents, for multiple errors that occurred during many decades of accumulated experience worldwide (throughout the United States and Canada, the United Kingdom and Europe, Africa and Asia). We have emphasized technological systems and used the data that are available from several key countries, most notably the United States.

*Table 5.1  Comparison and Summary of Derived Minimum Error Frequencies*

| Personal Risk | Minimum Error Frequency: 1 in Approximately . . . | Data Source(s) |
|---|---|---|
| Fatal air crash | 220,000 hours | World airlines 1970–2000 |
| Air event | 20,000 hours | U.K. occurrences 1990–1998 |
| Midair near miss | 200,000 hours | U.S., U.K., and Canada 1987–1999 |
| Fatal train crash | 20,000 hours | U.S. railroads 1975–1999 |
| Ship sinking | 350,000 hours | World shipping 1972–1997 |
| Auto accident | 6000 hours | U.S. data 1966–1998 |
| Industrial injury | 30,000 hours | U.S. data 1998 |
| ACSI injury | 200,000 hours | U.K. and U.S. data 1970–1997 |
| Medical error | 25,000 hours | U.S. data 1983–1993 |
| Homicide | 50,000 hours | Canada data 1973–1997 |
| Tire failure death | 25,000 hours | U.S. data 1991–2000 |
| Licensing error | 200,000 hours | U.K. and Canada data 1996–2000 |
| Boiler failure/defect | 175,000 hours | World data 1999–2000 |

Look at the numbers in Table 5.1: there is something we did not anticipate when we were working our way through all these numbers. We note that there is a remarkable, and at first sight perhaps an unexpected consistency in the error data, generally lying between 20,000 and 200,000 hours, except for auto accidents, and the apparent two groupings of the extremes of highest and lowest numbers. We have already explained that the higher rate of errors for auto accidents arises because of the high (random) density.

The lowest observed active error values are circa 200,000–350,000 hours for the ACSI, but are of the same order, and for others there is a clear bifurcation of risk by an order of magnitude (a factor of 10). So the figure we started with at the very beginning of this text (taken from auto fatalities in the United States) now looks like Figure 5.1.

The numerical values are derived from the error rate measures converted to rates per unit experience, and the accumulated experience is non-dimensionalized to the maximum observed. The initial (startup) rate is highly system and technology dependent and is qualitatively some 5 to 10 times larger than the minimum rate due to active errors. The minimum attainable rate is generally less than 1 in 100,000 hours based on the analysis and data to this point in history.

Why is the underlying (minimum) error rate of this value? Is there a deep-seated reason why the value is what it is? Is it a fundamental barrier or quantity? Or is it simply, perhaps haphazardly, a manifestation of our analysis and/or how humans interact with technological systems?

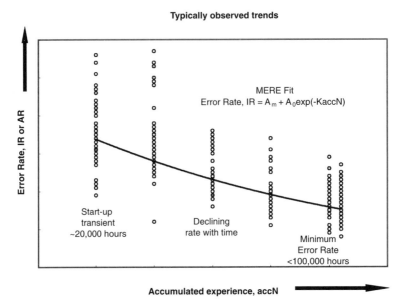

*Figure 5.1 Typical Universal Learning Curve.*


## *Equipment Failures and Super Safety*

The process of aging and failure is also well known in equipment. Many large modern technological systems (airplanes, oil rigs, power plants) utilize literally millions of components, which are all aging and wearing out. Maintenance of this equipment is essential.

There may well be other areas where such a learning process (curve) is operating and the concept is applicable. We have also found the same exponential trend in component failure data.

In his pioneering IEEE paper in 1964 on equipment failures, Duane showed that in tracking failure rate information it was best to plot it on an accumulated experience basis using the operating hours. He further argued that the failure rates were constant. After correcting for a reversed mislabeling of the two plots in his paper, we can plot his data using the DSM analysis approach, as shown in Figure 5.2.

It is then apparent that the rate is following an insufficient learning curve, not a constant rate, and that the IR is flattening out to a minimum after about 10,000 hours of operation. The MERE fit to these data conforms to this trend and is given by

$$IR \text{ (Failures per 1000 hours, F/kH)} = 7.4 + 138 \exp(-acckH/0.7)$$

**Aircraft Equipment Failure Learning Curve**

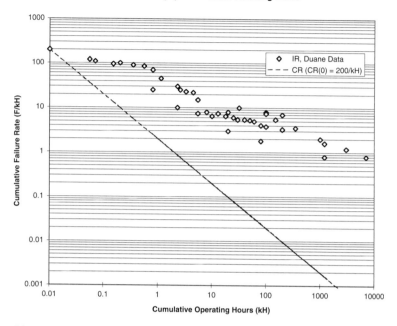

*Figure 5.2 Equipment failure data plotted as the IR and compared to the CR using operating hours for the accumulated experience. (Data source: Duane, 1964.)*

The implication is that there is indeed an embedded minimum failure rate in this particular equipment.

The failure interval is embedded when adding equipment into safety systems, duplicating and triplicating these systems for redundancy and diversity. Howard has studied low-failure-rate systems and calls this approach "super safety" by design. By tracking the 20th-century history of such multiple fail-safe architectures in aircraft, Howard showed these failures to possibly occur less than 1 in 100 million hours ($\sim 10^{-8}$ per hour). At this low level of mechanical and electrical equipment failure, they do not matter, and so-called common-mode (hidden) or human errors dominate. By our DSM data analysis, we have shown this minimum failure rate to be 1 in 100,000 hours ($\sim 10^{-5}$ per hour), or 1000 times higher, in modern technological systems.

Howard's suggestion in the search for "super safety" is to ultimately increase the level of automation, and hence the level of safety by design. We would also argue that the layering and adding of systems is not fruitful unless the technological learning curve is followed. *It is not possible to build completely reliable systems that we humans are likely to or can defeat.*

## 5.4 Speculations on the Active Minimum Error Rate

### Risk in Society

It could be that in modern society we are just starting to understand the concepts and the balance of risks and the measures of societal acceptance. The views of a country or nation on technological systems and their safety depend on whether these systems are perceived as alleviating hunger and providing for jobs and economic growth, or as a threat to the measure of wealth and comfort achieved through industrialization and technological systems. After all, humans have only been operating complex technological systems for some 200 years, starting with steam engines and factories in individual plants in the 19th century, and working up to computerized operation and control of large complexes in the 20th. This is a rich area for future psychological and socioeconomic study; we have concentrated on the error data to see what has been learned.

Since acceptance of risk by the individuals in a society is so highly colored by human perception, there is no absolute measure of liability, no assurance of minimization or existence of an international treaty. Each society, nation, state, and industry has its own regulation and laws, based largely on their own good or bad experience to date. There is no logical uniform standard, as Stephen Breyer (1993) has pointed out, even within a society.

Common sense says that we must still have accidents and errors in any system, and all we can do is reduce them as much as we can. This is the inevitability theory of "normal accidents," the pessimist's view of the world. This is the world of Perrow's "normal accidents," whereas the optimist would say if we just try harder and harder, we could perhaps eliminate the errors and thereby reduce their effects.

The DSM analysis indicates that we would expect everyone to learn from his or her mistakes (errors). From an initially high rate of errors when starting out with a technological system, we steadily reduce the rate by paying attention to safety and by having a learning attitude, correcting and counteracting our mistakes as we go. Thus we descend the Universal Learning Curve, the rate of decrease being some measure of how good we are at preventing accidents. The rate should be lower than the constant rate (CR), even for a mature activity. But errors still occur, and as we accumulate more and more experience, we find that the rate of decrease "levels out." We may even see an increase. This is when the minimum error rate has been reached—the nonzero finite error rate that cannot be eliminated, no matter how hard we try. *There are no "zero defects," just an attainable minimum.*

We now openly speculate and consider the two types of theory that might explain the presence and values of minimum error rates. Why is the minimum observed error rate not less than about 1 in 200,000 hours, a number that reappears time after time? Since we are dealing with the human machine, we cannot perform experiments on it or understand its complex inner workings in great detail. As we shall see, humans try to accommodate errors while machines try to eliminate them. For technological systems, we also try to distinguish *active errors*, where the human involvement is a primary or dominant contribution, from *passive errors*, which are due to lack of involvement and action. We have speculated already that learning is "active" and forgetting "passive," so the actual error reduction rate depends on the learning-to-forgetting ratio.

## Making Errors and Taking Risk

Now, we have also shown that serious errors of judgment occur in key decision making, especially when planning and dealing with complex technological systems and social situations. The *confluence of factors* may seem unrelated but can include unexpected and often hidden interdependencies that contribute to the risk exposure.

So why do we not recognize and correct for them?

The first theories are observational and psychologically causative in nature: the errors are due to problems of perception and solution. Just as in medicine, there are many descriptions of the symptoms exhibited by errors and their rationale. Thus, there are detailed discussions of error types, attributes, probabilities, causes, and distinctive features. Errors are classified as to whether they are errors of omission or commission, in a mechanistic manner. In his fundamental model, Rasmussen (1986) proposes failures of the skill, rule, or knowledge base that are due to some malformed or misinterpreted mental process model. Errors are thus due mainly to psychological and interpretative causes. But none of these "models" gives a quantified value for the underlying error rate, just good reasons for the existence of errors.

One approach is to try to understand why the human errors were made in the first place. Based on qualitative observations using simulations, the cognitive psychologist Dorner (1996) suggested that the mistakes we make are due to the four characteristics of complicated systems. There is the complexity of the system interactions; the dynamic time dependency of the process relationships; their "intransparence" or lack of visibility to us; and our ignorance and mistaken hypotheses due to having a wrong or incomplete mental model of the system. These are clearly contributing factors in the confluence that occurs when errors are made. All the experimental simulations analyzed used simple elapsed time as a measure for the accumulated experience and, as we now know, should really be reinterpreted on the basis of a learning model.

Both the DSM and the data we have analyzed show that the learning and error reduction process is inherently nonlinear and that the measure for gaining the needed experience is not necessarily calendar time. Basically, as humans, we prefer to think in static terms, and we usually make built-in linear assumptions about the things we observe and how they might develop. We are quite bad at dealing with the realities of the inherently obscure or complex and at disentangling the nonlinear trends and the often invisible interrelationships. We have to learn how any system behaves and responds by trial and error, making mistakes as we go, and then, we hope, embody the correct knowledge in the needed training and mental models.

Thus we may only qualitatively understand what happened. We can establish what factors and influences, and which human and situational attributes or misperceptions, contributed to the error. This is similar to the inquiries into the causes of all those accidents that could not happen, won't happen to me, or certainly can't happen again. . . .

Except that they might!

The psychological and sociological literature is full of valuable analyses of human behavior and qualitative "models" to explain the observed behaviors. Some descriptive theories have been developed explicitly to analyze errors and to provide a rational basis for the reasons why and how errors and accidents occur. These many and varied models include the hypothesis that there is "behavioral compensation" (what we would call learning) when there are incentives or motivations for changes, be the consequences physical or mental rewards, penalties, and/or pain and suffering. People make mistakes because they take risks; the perceived or actual outcome of the risks can influence the subsequent behavior.

To compensate for lack of understanding, we humans may adopt a convenient ploy and ignore and deliberately avoid the truth by not fully considering the consequences of our actions. We take risks. The best decision makers may indeed be like Napoleon Bonaparte: first acting quickly with a risky "trial selection" and then intuitively or logically managing the emerging consequences using an acute sense of situational awareness.

*Thus, the theories say, accidents and errors are the result of risks taken, for whatever reason, whether knowingly or not.* The literature is full of elegant names and there are many qualitative theories. For example, so-called Risk Homeostasis and/or Risk Motivation Theories examine the effects of many factors on risk-taking, trying to distinguish between the different types, perceptions, consequences, and goals that determine risk-taking behavior. It is also postulated that there is explicitly or implicitly a "target level" of risk, which can be dynamically varying. Continual mental feedback from risk perception occurs, shifting the actions and targets as they lead to more or less risk. We would also call this observed behavior "learning"—in this case, from one's mistakes.

### Risk Theory Validates the MERE

Our observations and conclusions of error reduction by learning and the existence of an irreducible minimum error rate are consistent with and supported by risk-based theories. Manfred Trimtop succinctly states:

> This points to a possible learning effect that carries over into subsequent behavior. . . . The data suggest that target levels of risk are constantly re-assessed, or even formed, when new experiences are being made. . . . But, the results . . . also indicate that no matter what the rewards and penalties are, most people will still take risks. (Trimtop, 1990)

Risk Theory has been applied to explain trends in traffic accidents. The data, says Professor Gerald Wilde (1989), "support the notion that the accident rate is a function of the target level of risk." But there are no numbers given for what all this means, or what actual targets are formed. Risk Theory is a qualitative psychological explanation or model, not a precise mathematical and physical law or relation.

But we have provided a quantitative analytical basis for these purely empirical observations by solving the minimum error rate equation. We now know that the accident rate, A, at any accumulated experience N* is very nearly given by the learning curve:

$$A = A_M + A_0 \exp(-N^*)$$

or

$$A = f(A_M, A_0, N^*)$$

In words, this mathematical relationship states simply: "The accident rate is a function of the initial rate, the minimum attainable rate, and the accumulated experience."

The MERE result explains and quantifies the observational hypothesis embedded in the risk-based theories, if we assume that the minimum attainable rate, $A_M$, is actually equivalent to the "target rate." After all, we would surely want to reduce errors to the minimum consistent with our perception of the risk and the progress toward it, which may of course both differ from individual to individual or organization to organization. The instantaneous goal in any given situation at any experience level is thus taken as $A_M$, the *perceived or apparent* attainable minimum rate. The perceived rate of error reduction is the fractional or actual reduction from the initial error rate, $A/A_0$. The motivation for action is the rate at which errors are perceived to be reduced, which decreases exponentially as experience is gained. The MERE model and the mathematical form are exact and can now be restated in the words of "Risk Theory" as: "The accident rate is a function of the initial rate, the target (minimum) rate, and the accumulated experience."

We have avoided the debate on error causation up to this point, having so far described the observed errors as being somehow embedded in the overall human–system interaction. Thus, we assume that the minimum rate is an estimate—however crude—of the underlying overall and mostly human error rate.

### *Mechanistic Models: Quantifying the Minimum Error Rate*

The second type of theory is openly mechanistic in nature: the errors are caused by and within the decision process itself.

In his pioneering work on automata, the gifted mathematician and co-inventor of the modern computer John von Neumann examined automatic computers, or self-regulating and self-replicating machines. Here the analogy of computers to humans is obvious. Both are making complex calculations and decisions internally, while we observe the results appearing externally. The myriad of calculations and number crunching constitutes a vast and essentially "incomprehensible organism," according to John Newman. Internal feedback occurs inside, correcting the answers in an analogy to closed-loop control systems. Truly here is the ultimate human–machine interface, with the human itself acting also as a machine via the functioning of its own brain and central nervous system (CNS).

Von Neumann, along with Alan Turing, draws an analogy between the living organism and the computing machine, while recognizing that analogy to be imperfect. Thus, nerve (so-called neuron) operation in the brain obviously is not exactly the same as the digital binary switch operation in a computer. Von Neumann observes that the CNS has about 10,000,000,000 neurons ($10^{10}$) or more and performs about 100,000,000,000,000,000,000 ($10^{20}$) nerve operations in a lifetime.

Von Neumann envisages exploration of space using self-replicating automata, which we could program or instruct to do so out in space. After launching, this would give a vast number of these so-called self-replicating "von Neumann probes" searching space far more efficiently and quickly than just by sending humans.

But the limitation on self-replicating automata is the problem of errors in the replication process or in the instruction set. To quote von Neumann from his "General and Logical Theory of Automata," where he discusses the difference between error minimization (learning) in humans versus error isolation (diagnosis) in machines:

> It is unlikely that we can construct automata of a much higher complexity than . . . we now have, without possessing a very advanced theory. A simple manifestation . . . is our present relation to error checking. In living organisms malfunctions occur. The organism has a way to detect them and render them harmless . . . to minimise the effect of the errors. In artificial automata (computers and machines) every attempt is made to isolate the error as rapidly as feasible. *As soon as the possibility exists that the machine may contain several faults . . . error diagnosing becomes an increasingly hopeless task.* (Emphasis added)

The problem is the uncorrected self-replication of multiple undetected and uncorrected errors, exactly the issue we have identified in the error causes of accidents and events of all kinds.

This is also a fundamental question in the debate on the origins and possible unique-ness of human life itself and whether replication is error free. The human brain, accord-ing to Barrow and Tipler in *The Anthropic Cosmological Principle*, stores about $10^{10}$ to $10^{15}$ bits (one bit = one cell) of information and between each cell can fire or cycle neurons at a rate of about 100 per second. For comparison, large computers store $10^{10}$ bits of information and can perform gigabytes of digital (binary) operations a second (1,000,000,000 or $10^{9}$), which is much slower than all the neurons firing ($10^{12}$ to $10^{17}$). Now it takes about $10^{9}$ operations (nerve firings) for even simple human tasks, which means a processing rate of about $10^{11}$ flops per second (or ~$10^{2}$ Gfp/s).

We have found a minimum error rate estimate of the order of 1 in 200,000 hours, or about 1 in $10^9$ seconds. Thus, a simple human task has a potential error rate of order 1 in $10^{11}$ times $10^9$ or 1 in $\sim 10^{20}$ flops. We correct for these all the time. Now if a human error occurs during many gigaflops (an action, an idea, a motor activity), which take, say, 1 minute, that is about $10^{10}$ times $10^2$ or about $10^{12}$ firings, and our error rate in neurological terms is then of order 1 in 100,000,000 neuron operations ($\sim 10^{-8}$). This does not seem an unduly large number.

We do not know if this estimate of 1 error in 100 million brain operations is correct, or how significant it is. To put it another way, it is like having a 1% chance of error for every million computer operations, which seems a not unreasonable number. Are we at the limit of the ability of the human brain for error-free information processing? We just do not know.

## 5.5 UNIVERSAL LEARNING CURVES: VALIDATING THE THEORY

The entire DSM analysis has reduced the many data sets to a simplified but highly convenient and uniform learning-curve form. The theory provides a compact notation for the exponential model, which we termed the MERE, which relates the error rate to the accumulated experience.

*Recall that the fundamental hypothesis for constructing the learning curve is that the rate of learning (rate of error reduction) is directly proportional to the rate of errors actually being made.* Thus, we are learning from our mistakes.

For an initial error rate, $A_0$, and a final minimum error rate, $A_M$, the form of the MERE is given in Appendix A as

$$(1 - A/A_M) = (1 - A_0/A_M) \exp(-N^*)$$

where $N^*$ is a measure of the experience.

So the two limits are:

$A = A_0$ the initial rate for $N^* \rightarrow 0$, at the beginning with no experience, and
$A = A_M$ the minimum rate as $N^* \rightarrow ^\circ$ , in the limit of the maximum
     possible experience

The result can be written in handy nondimensional notation as the exponential relation

$$A^* = A_0^* \exp(-N^*)$$

or, as a nondimensional error, E*,

$$E^* = \left( \frac{A^*}{A_0^*} \right) = \exp -N^*$$

where

$A^* = (1 - A/A_M)$
$A_0^* = (1 - A_0/A_M)$
$N^* = N/N_{max}$, the nondimensional accumulated experience

These results can be generalized somewhat by including the learning rate constant, k, as follows:

$$A^* = A_0^* \exp(-kN^*)$$

and the nondimensional error,

$$E^* = \exp(-kN^*)$$

where a positive k-value is learning and a negative k-value is forgetting.

To compare our disparate error data and MERE results from the many diverse sources, we can reduce all the data to the above nondimensional form. The accumulated experience is divided by the maximum experience, and the error rates by the initial and minimum values.

We have derived the MERE for many cases, as shown in the text using the simplified form

$$A = A_M + A_0 \exp(-N)$$

Or, in words, the accident, error, or event rate, A, falls away exponentially from the initial rate, $A_0$, toward the minimum attainable, $A_M$, as experience is gained.

We can simply use the values we have estimated for $A_0$ and $A_M$, and calculate the E* value for each actual data point A-value and plot it against the N*. We illustrate the results of such an analysis in Figure 5.3, which includes data for deaths from diseases, rail, autos, and recreational boats, fatal airline crashes and near misses, tire failures, mining accidents, and latent errors. *We believe this is the first time ever that data from such disparate sources and human activities have been plotted and shown on one graph.*

Also plotted is the theory predictions for E* = exp(–kN*). We certainly expect the data to follow this form, which it does, but appears to be a little low for k = 1 compared to some of the data shown. Thus, it must be that k > 1, which is not at all surprising. The comparisons are sensitive to the values derived for $A_0$ and $A_M$; also, not all the activities have full data or a fully formed learning curve; and the choice of the initial values and

**Universal Nondimensional Learning Curve**

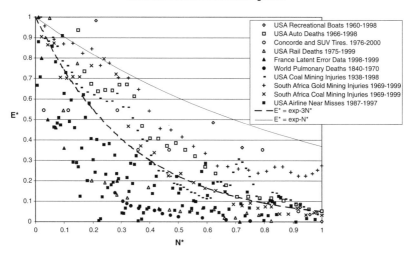

*Figure 5.3 An example of the Universal Learning Curve for a wide range of available data.*

the accumulated experience is predicated on what data are actually available. In addition there has been no *a priori* fitting or adjusting of values, data, or constants to obtain this level of agreement.

Adjusting or allowing for different learning rate constants, k, and renormalizing to the accumulated experience, N, when the minimum error rate was actually reached will line up the theory curve and the data, as shown by the example using a value of k = 3.

*Therefore, we regard this trial attempt to unify the world's data using the DSM as very encouraging and validating both the theoretical and data analysis approach.*

Of course, we can improve the fit by some minor adjustments, particularly using the learning rate constant, k, which is given in Figure 5.3 by the range of 1 < k < 3. The best fits to these typical data sets are shown in Figures 5.4 and 5.5.

For the full MERE we expect the exponential model to work, and the fit shown is

$$E^* = 1.0 \exp(-3N^*)$$

which implies a learning rate constant of k ~ 3. The linear (first order) approximation is given by

$$E^* = 0.96 - 1.04N^*$$

These values of 0.96 and 1.04 are extremely close to the expected value of unity (1) that the theory suggests, and certainly within the errors due to the data scatter. This is a satisfying result in the sense that a residual or minimum error still remains!

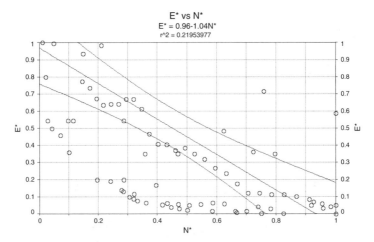

*Figure 5.4 The typical linear MERE approximation to data, including the 95% confidence limits.*

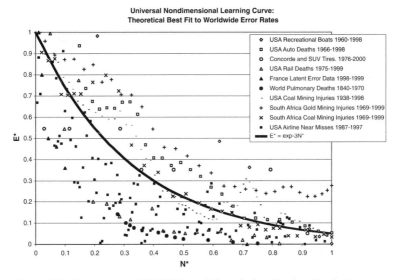

*Figure 5.5 The exponential MERE model prediction fitted to the data.*

## 5.6 ERRORS AND TECHNOLOGY: ARE YOU SAFE?

### *The Journey's End*

Every day we all read and hear of new disasters, we see human deaths and injuries in the media and headlines, and we are told of new risks, fears, and issues.

For us, it has been a long journey into a world full of tragedy, accidents, and errors. Our lengthy and, we believe, unique investigation into errors in complex technologies and recreational and societal activities has led us to major conclusions. We conclude that there are relationships between diverse activities in respect to learning curves and apparently irreducible minimum error rates. We have developed a formula, which can be applied to data, complete technologies, systems, and individual events. The rate at which errors are reduced through learning depends on the number of errors being made. We have shown that minimum error rates can be deduced and predictions can be made for the future.

Our voyage has been through 200 years of technology in multiple industries and many spheres of human activity. Our data analysis has covered more 200 million aircraft flights, hundreds of millions of medical patients, some 1 million ship-years, more than 200 million workers, thousands of millions of vehicle miles, thousands of millions of train miles, millions of components, hundreds of years of mining, and thousands of homicides and accidents. This vast array of information is for multiple errors that have occurred during many decades of accumulated experience worldwide. We have empha-sized technological systems and used the data that are available from many key coun-tries, most notably the United States, and from the continents of Europe, America, Australia, Asia and Africa—but the journey is not yet over.

*The crucial difference in our approach from the standard methods of recording and plotting such event data is our use of accumulated experience to determine the learn-ing curves. We have been able to derive a Universal Learning Curve as a result of the analysis of a great amount of event data.*

We believe our extensive collection and analysis of real data to be unique, and the results to be entirely new.

### The Global Trends

We were pleasantly surprised to find a wealth of detailed error data on some specific government agency websites, additional to those reports from the major industries and specific interest groups. The World Wide Web provides unrivalled access to much information in many countries. However, it is quite apparent that much more could and should be done to provide more error data worldwide, so that detailed analysis can be conducted. Indeed, we found that some professions are still reluctant to make data pub-licly available (e.g., the medical profession). Perhaps in some cases they just don't keep such records, or the data are inaccessible. Lack of public data means reduced learning capability, likewise fostering internal "closed" reporting systems, which are only useful to the company(ies) concerned.

As technology spreads, safety and risk are without frontiers. Errors do not respect national boundaries. Globalization of technology demands much more event data on incidents as well as accidents. In order to progress corrective actions intended to reduce the number of serious events, some industries have international organizations that develop global recommended standards (e.g., the International Civil Aviation Organisa-tion, ICAO; the International Maritime Organization, IMO; and the World Association of Nuclear Operators, WANO). If we are truly global, then we must have standards that are applied worldwide, including the recording and the dissemination of data.

We live in a technologically stimulating environment and have made great strides in improving safety over the past 30 or more years, especially considering the rate of technological change. Accident/incident investigations show that the vast majority of events are the result of simple errors, mostly human; it is rarely the technology alone that causes the grief. The events have differing *confluences of circumstances and events*, which makes them increasingly difficult to anticipate. However, there are some errors, such as those stemming from common modes, which can be addressed by design.

Considerable effort to reduce simple human errors is being carried out in the major technological industries. But there is a need to significantly improve the acceptance of the factors involving human errors and the actions required to include acceptance by corporate and senior management. To reduce their liability exposure, executives and board members must ensure and show that they are being prudent. They must have continuous improvement in performance and support a learning environment that tolerates mistakes and enhances safety as well as profits.

### *The Stubborn Minimum*

We found from the data that the minimum frequencies for active errors in technological systems were about 1 in 200,000 hours and about one order of magnitude less for more passive systems, such as dams, which are subject to natural vagaries and forces. Error rates in major technological industries are proving very stubborn to improve, as they appear to have reached an asymptotic value. There are several *influencing factors*:

- Increasing pressure is placed on operations in a growing world economy, particularly transport, producing gridlocked systems, which in itself is likely to increase the number of events if the error rate is not reduced.
- The competitive element is increasing, with the drive for increased efficiency and profits in a privatized global marketplace.
- Business is changing rapidly, and it is evident that technical expertise is nowadays often excluded from the boardrooms. Technical expertise is on the decline in some industries and stretched, often causing great competition for scarce technical resources.
- The short-term management approach does not help long-term career planning and can lead to an unsettled working environment.
- The development of technological innovation is stretched to cope with expanding business and the fast pace of change, which can affect both the adequacy and coordination of research and development.
- Public awareness and reduced tolerance for errors resulting in what are seen as unnecessary deaths and injuries produce a perception from skeptical public elements that safety is being traded for economics and dividends.

- The extensive subcontracting of important functions (e.g., maintenance) can lead to a reduction in effective monitoring of safety standards, which is a crucial element in controlling safety.
- Increasing litigation can have a negative effect on improving the learning environment when people are reluctant to admit to their errors for fear of retribution or prosecution.
- Human errors are the crucial element, and the societal influences need to be further reviewed and addressed.
- In the so-called almost completely safe industries, where accidents have become rare events, the need to improve the error rate is heavily dependent on creating and sustaining a true learning environment.

It is a truism that there is no simple answer to the reduction of human errors. The differing *confluence of factors* tells us that, all too often and in graphic detail, human beings and societal interactions are complex.

We are indeed up against a fundamental barrier to error reduction. It may well be that we must take action. We must either fundamentally change the way we use and manage technology; or find new and better systems to help us in measuring, monitoring, managing, and regulating our progress on our journey along the technological learning curve.

### The Future Learning Curve

We like to think positively and focus on what we have learned. Our investigations have shown that we humans have learned from our mistakes in the past and we expect this to continue in the future. However, it is becoming increasingly difficult and complex to anticipate event sequences. We do not believe that zero accidents are attainable, as shown by our analyses, although we may see extended periods of no accidents in a particular industry, technology, or operation.

We believe the answer to the question "How safe are you?" depends, to a large extent, on the individual's contribution to the learning environment, *in a given role and circumstances.* Only then are we able to reduce human errors to the minimum in our technologically driven world. *We also need to ensure that we remain in quantifiable and measurable control of our technological environment.*

We have shown that errors in modern technology are real and challenge us all. The risks are measurable and manageable. We have shown that a Universal Learning Curve exists; all we have to do now is to actually follow it. *We must learn how to learn.*

# Appendix A

## UNIVERSAL LEARNING CURVES AND THE EXPONENTIAL MODEL FOR ERRORS, INCIDENTS, AND ACCIDENTS

### SUMMARY

A simplified and useful theory is derived for the failure rate of systems, components, or technologies, including the effects of human error. The analyses for the failure rate and error probability are based on the existence of a finite minimum rate. The minimum error rate equation (MERE) is derived and presented. This equation leads directly to nondimensional representations and parameters, which use the maximum accumulated experience and the minimum (asymptotic) error or accident rate as the normalizing parameters.

As a direct result, a Universal Learning Curve (ULC) is derived, and apparently quite independent error data sets are correlated for a wide range of human and worldwide activities. Since human error is the key unifying parameter, the analysis also provides a basis for the estimation and prediction of error rates, both now and in the future.

The implication for the search for "zero defects" is that this is not attainable. Instead, error and defect reduction occurs as learning increases with increased accumulated experience, until the attainable minimum is reached.

Application of the approach to errors, accidents, incidents, misdiagnoses, and misadministrations is described. The new approach (called the Duffey–Saull Method, DSM) has been examined as to how it may be applied to transportation (e.g., rail, road, air, and shipping), industry (e.g., manufacturing, chemical, nuclear, and mining), and other common technological and sociological systems (e.g., medical and medication error, and the insurance and prediction of risk).

The trends in the available accident, failure, incident, error, or injury data can be represented universally as exponentially falling away with time (or increasing experience), to an apparently irreducible minimum. Error rate theory provides a theoretical basis, which explains the observed trend and takes the form of a classic learning curve. The influence of forgetting is to reduce the downward trend, the reduction depending on the relative rates of the Forgetting to Learning rate constants.

## ERROR RATES

The error history, as measured by accident, injury, or morbidity rates, approximates a simple exponential curve. The standard formulation for the probability of failure, $F(t)$, is then given by

$$F(t) = \int_0^t f(t)\,dt$$

$$F(t) = 1 - e^{-\int_0^t h(t)dt}$$

where $h(t)$ is the hazard function. In terms of the failure probability, $p(t)$, and assuming a constant or suitably time-averaged failure rate, $\lambda$, then we have the standard relationship,

$$p(t) = F(t) = 1 - e^{-\lambda t} \tag{1}$$

Now for a finite (nonzero) asymptotic minimum failure rate, $\lambda^*$, we must revise this expression. The failure rate is effectively a time-averaged summation over all known and unknown components or contributions due to design, quality, manufacturing, operation, maintenance, and wear, including any human error contribution. In other terminology, these multiple "barriers" to failure are being penetrated by design, manufacturing, maintenance, procedural, and operational errors. Thus, formally we have,

$$\lambda = \sum_i \lambda_i$$

where the summation is over all failure modes and types, both known and unknown. Usually, the failure rate is assumed to fall to zero at long times, but with the existence of a finite minimum rate we may write

$$\lambda - \lambda^* = \sum_i (\lambda_i - \lambda^*)$$

Thus the failure rate equation becomes

$$p(t) = F(t) = 1 - e^{-(\lambda - \lambda^*)t} \tag{2}$$

For rare events with small failure rates, $(\lambda - \lambda^*) \ll 1$, and we may expand the exponential to give, to first order,

$$p(t) \approx (\lambda - \lambda^*)t \tag{3}$$

or

$$p(t) \approx \lambda^*(\lambda/\lambda^* - 1)t$$

## Minimum Error Rate Theory and Learning

The fundamental assumption is that the rate of error reduction (learning) is proportional to the number of errors being made. *The rate of decrease in error (event) occurrence is proportional to the event rate.*

For the instantaneous rate of change of failure probability, the form of the equation describing the failure rate is given by using the analogous classic formulation from failure rate modeling. If we are learning, then the rate at which errors occur is indeed directly dependent on how many errors exist. Thus, taking the rate of change of the failure rate as proportional to the instantaneous failure rate, then, for any *interval of accumulated experience*, $\tau$,

$$\frac{d\lambda}{d\tau} \propto \lambda(\tau) = -k\lambda \tag{4}$$

where k is a constant, which we refer to as the "learning rate constant," being the characteristic measure for a given technological system.

So from (4), we have the equivalent differential form of the minimum error (failure) rate equation (denoted as DMERE):

$$\frac{d\lambda}{d\tau} + k(\lambda - \lambda^*) = 0 \tag{5}$$

The rate of decrease of failures (learning) is proportional to the number of failures (errors). We equate failures with errors. Previous equations and derivations by others did not explicitly include the lower or asymptotic rate, and hence *implicitly* assumed that $\lambda^* = 0$. Prior work also does not distinguish clearly the accumulated experience interval, $\tau$, from the normally used elapsed or arbitrary clock time, t.

## ACCUMULATED EXPERIENCE AND THE EXPONENTIAL MODEL

In any activity, as humans we accumulate operating hours, training time, and days or hours worked. But for our technological or social systems, we accumulate the numbers of equipment operating, distances traveled, pieces manufactured, patients diagnosed or treated, plants operated, computing time, or productive output. These are all different measures, but always include the human component somewhere, whether explicitly or implicitly, be it in the design, licensing, operation, manufacturing, training, mainte- nance, repair, supervision, management, or procedures adopted.

Therefore, for any given technological system or human activity, the elapsed time, $\tau$, is really the accumulated experience with that system, and *not* just the arbitrary or counted number of hours, days, months, or years passing. Unfortunately, since this is convenient for record keeping and reporting, elapsed time is the usual means by which we humans record accidents and errors (e.g., by calendar month or by fiscal or calendar year).

For a general *accumulated experience* interval or measure we have $\tau = T$, with the hypothesis that there is a minimum attainable failure rate, $\lambda^*$. Integrating (5) from some beginning at $\tau = 0$ to $\tau = T$, we obtain the solution as the relation

$$(\lambda - \lambda^*) = (\lambda_0 - \lambda^*)\exp - kT$$

$$(1 - \lambda/\lambda^*) = (1 - \lambda_0/\lambda^*)\exp - kT$$

(6)

where $\lambda_0$ is the initial failure rate value with little or no experience ($kT = 0$), and for large experience ($kT \rightarrow °$ ), then $\lambda \rightarrow \lambda^*$, the asymptotic or minimum error rate.

We note that equivalently, therefore, the failure probability is:

$$p = p_0 \exp(-kT)$$

(7)

In terms of observed accident or error rates, A, per unit interval as usually reported in data files, then $A \propto \lambda$, and equation (5) is written as the minimum error rate equation, MERE:

$$\frac{dA}{d\tau} + k(A - A_M) = 0$$

(MERE)

Thus we have the solution to the MERE written as

$$(1 - A/A_M) = (1 - A_0/A_M) \exp(-kT)$$

(8)

where $A_M$ is the minimum error rate, and $A_0$ the initial error rate or for no experience.

In estimating the accumulated experience, this is in practice related to the total amount of recorded or accumulated experience, $N_T$. In modern settings and when recorded, this is typically measured by items such as the system operating time, observed intervals in a study, total airline flights, shipping years sailed, items manufactured or observed, distance traveled, number of vehicles moving, or industrial working hours.

Thus, we may nondimensionalize the actual accumulated experience, $N(\tau)$, of any experience interval, $\tau$, by writing the characteristic timescale $kT$ in terms of the fraction of the total accumulated experience, $N_T$ or accN, as follows:

$$kT = kN(\tau)/N_T = N^*, \text{ say},$$

where $N_T = \Sigma N$, and $N^* \to 0$ for no experience.

Hence (8) becomes

$$(1 - A/A_M) = (1 - A_0/A_M) \exp(-N^*) \tag{9}$$

so that as $N^* \to {}^\circ$ then $A \to A_M$, *the asymptotic minimum error rate.*

This is the fundamental useful result of the analysis. Equation (9) will be termed the *minimum error rate equation* (MERE) and is the so-called *exponential model* for the error rate.

To a good approximation, $A_0/A_M \gg 1$, and we may write (9) in the simple and convenient MERE approximate form:

$$A = A_M + A_0 \exp(-N^*) \tag{10}$$

We use this approximate MERE form throughout for fitting the trends in the IR and AR data, as it is the simplest and most robust model. The limits are retained of:

$A = A_0$ the initial rate for $N^* \to 0$, at the beginning with no experience
$A = A_M$ the minimum rate as $N^* \to {}^\circ$ , in the limit of the maximum possible experience

A plot of the instantaneous observed accident or error rate, $A$, normalized by the initial and minimum rates, versus the nondimensional accumulated experience, $N^*$, should therefore exhibit a simple exponential decline with increasing accumulated experience. We call this the Universal Learning Curve (ULC), and it would be followed for any system that is well behaved in its asymptotic limit and as experience is accumulated.

From the MERE (9) and (10), we can also expect any new system to have an initially high error rate, called a startup transient. In the presence and enforcement of suitable safety management and other systems, the rate will then trend exponentially steadily downward as experience is accumulated. Thus, we can construct and use such a curve (the ULC) to determine if the data or system is well behaved and is tending toward or has attained its minimum error rate.

## RENORMALIZED NONDIMENSIONAL FORMS
## AND THE LINEAR APPROXIMATION

We give here some useful practical expressions. Now the MERE (9) can be written in convenient nondimensional form as

$$A^* = A_0^* \exp(-N^*) \tag{11}$$

where $A^* = (1 - A/A_M)$ and $A_0^* = (1 - A_0/A_M)$.

The asymptotic failure rate is still given by $A_M$ as $N \to \circ$ , and the asymptotic (Beysian) meantime between accidents, $\Phi$, is then $1/A_M$.

We may now write the failure rate equation in an even more compact normalized form. By defining a nondimensional failure or error function, $E^*$, equation (10) becomes

$$E^* = \left( \frac{A^*}{A_0^*} \right) = \exp - N^* \tag{12}$$

The expansion of the exponential term gives, of course,

$$E^* \sim 1 - N^* + \cdots \quad \text{or} \quad 1 - E^* \sim N^* \tag{13}$$

Thus, to first order only, the trend of $E^*$ with $N$ or $N^*$ may seem to be nearly linear with a negative slope.

For the case when $A_M \ll A_0$, we have the approximate result we have given earlier,

$$A^* \sim \exp(-N^*) \quad \text{or} \quad \ln A^* \sim -N^* \tag{14}$$

with the linear approximation

$$A^* \sim 1 - N^* \quad \text{or} \quad 1 - A^* \sim N^* \tag{15}$$

### Constant Error or Failure Rate: Special Cases and Limits

In many systems, we may have a constant workforce or manufacturing capability, within which observations are made at regular and unchanging intervals, or averaged over the same time frame. The error rate may be trending down or up or may be nearly constant.

Consider the special limit of when the error rate is nearly constant, having attained a minimum, or is remaining simply at a constant value with monotonically increasing time or accumulated experience. Such a situation could easily occur with a constant fleet size and no improvement or reduction in error rates.

From (9) and (12) we have the MERE that describes the ULC:

$$(1 - A/A_M) = (1 - A_0/A_M) \exp(-N^*) \tag{9}$$

or

$$E^* = \left( \frac{A^*}{A_0^*} \right) = \exp - N^* \tag{12}$$

Now if we observed for m reporting intervals (such as years or months, which are the usual or conventionally chosen intervals), and there are a constant amount or number of N accumulated experiences, each interval (as is true for, say, the nearly same equipment or fleet size, and no large or new system changes) for m intervals, then

$$\begin{aligned} N^* &= N/N_T \\ &= N/mN = 1/m \end{aligned} \tag{16}$$

Substituting (12) into (16), we obtain the Universal Learning Constant-rate Curve (ULCC),

$$E^* = \exp(-1/m) \tag{17}$$

Thus, as we accumulate a large experience base, $m \rightarrow °$, and we have $E^* \rightarrow 0$, and $A \rightarrow A_0$, the same (initial) value at all times. This is a constant rate, CR. However, for $m \gg 1$, $E^*$ for any CR will appear to fall away as (1/m), and if it rises, it must be because a minimum error rate has not been reached.

Two other interesting limits exist for the ULCC:

(a) No change in error rate, observation basis, or reporting interval: for $A_M \ll A_0$, which could be the case after some initial learning and experience has occurred, and a constant error rate, CR, is observed for m reporting intervals, then (17) becomes,

$$\begin{aligned} (1 - A/A_M) &= \exp(-N^*) \\ &= \exp(-1/m) \end{aligned} \tag{17}$$

which to first order is

$$= 1 - 1/m - \cdots$$

Thus, the apparent minimum rate is simply steadily decreasing inversely with time as the observed rate divided by the number, m, of accumulated or reporting intervals. The observed error or accident rate appears to be

$$A \sim A_M/m \qquad (18)$$

so that as $m \to {}^\circ$ , then $A \to 0$.

(b) No decrease in error rate: for $A = A_0$ or $CR(0)$, the initial observed or recorded rate always exists, depending on when observation started and the error rates were initially recorded or observed.

Since we have that $A = A_0$, then $A_M$ has the apparent rate of $A_0/m$.

Alternatively, from (15) and (18),

$$N^* \sim A/A_M \qquad (19)$$

These limiting cases provide a grid or network of constant rates, CR, which is traversed by the actual data, overlaid with the reality of varying rates, AR, as experience is gained and time passes, and procedures, systems, and reporting change.

A useful form is to write the constant rate as the identity

$$CR(N) = CR(0) \ (N_0/accN) \qquad (20)$$

where $N_0$ is the initial experience: and $CR(N)$ the expected rate based on the accumulated experience, accN, up to that point.

Ideally, one would like an invariant system of measuring and reporting, spread over a large history to acquire sufficient trend information. In reality, this is the exception, as the apparently never-ending quest for modernization, improvement, and change continues.

## PRACTICAL EXAMPLE OF LEARNING CURVES FOR ACCIDENT AND ERROR RATES USING THE DSM

The Duffey–Saull Method (DSM) is outlined below as the preferred approach. Failure probability theory clearly indicates that the reported accident or error data should always be plotted according to the forms suggested by equations (8)–(14). We generally use equation (10) to determine the initial and final or minimum rates.

The available or reported data generally do not have failure rates but report fatal accident, error, injury or industrial accident numbers, or rates based on observed or recorded time intervals of experience.

Therefore, we construct the error or accident rates analogous to equations (8)–(14) simply by redefining the accident, injury, or error rate per unit instantaneous experience (IR) or accumulated experience (AR). The constant rate trend is usually derived from equation (20), with the originally observed or estimated initial rate, CR(0).

We do not suggest using calendar years or times for reporting intervals for two reasons. First, the year time is an arbitrary interval for a technological system and increasingly so for humans (but not for funding or fiscal reporting purposes, which is perhaps why it is most widely used!).

Second, the accumulated experience (hours worked or risk exposure time) may well vary year-by-year or reporting interval, as the number of systems, people involved, or actual operating time changes. The preferred parameter is the *accumulated experience*, however it may be measured or defined (flights flown, surgeries performed, diagnoses made, decisions made, calculations or coding written, objects manufactured, hours worked, etc.).

The usual accident and statistical reporting schemes are based on calendar years. This yearly time scale is not useful in a general sense, other than to justify programs and funding, or to make the data understandable to those who think in terms of years passing, which is, of course, a purely natural human attribute. We recast the data into Excel format tables in IR, AR, and CR values versus the chosen accumulated experience, as shown by the examples in the text.

We do not know *a priori* the value of the asymptotic minimum rate, $A_M$, or *ab initio* if it has been attained or approached at any time. From the data, we may take the observed lowest value as the best estimate of the minimum rate, $A_M$, and we can also see if the curve of E* is trending toward a minimum in a well-behaved exponential fashion.

*If the AR exceeds the CR, this is a clear signal of insufficient learning.*

To fit the data, we may and do use available statistical methods. These include the least-squares and other residual curve fit options in commercial software, such as the exponential equations available in the software programs like TableCurve 2D, or in other computerized routines. Confidence limits for the fits at suitable levels (one sigma, two sigma, and 90%, 95%, and 99%) can be made for both the historic data and for predictive purposes, using both the instantaneous (IR) and the accumulated (AR) rates.

The MERE form given by equation (10) is the most useful approximation, and the estimated values are shown for the specific cases discussed throughout the text.

To place the minimum rate number in perspective, we note that the minimum rate contains contributions from all the possible causes, be it operator, maintenance, mechanical, or control errors. Each level (or "barrier") of defense introduced or used against failure due to error presumably provides a reduction in the instantaneous rate.

Hence there is or should be a trend down the Learning Curve toward the minimum achievable rate. The performance-enhancing methods introduced or adopted can be many, such as training, guidelines, maintenance, control, or even self-preservation. The evidence suggests that humans throughout the world exercise a high degree of self-preservation, adopting a high order of skill and obtaining a low order of error, but that the minimum achievable error rate is still finite (nonzero). *There are therefore no zero defects according to the minimum error rate theory and according to the entire available human experience database.*

Mathematically and physically, we cannot simply assume that $A_M = 0$ if the data do not agree. This finite minimum presumably exists because humans are involved, despite the designers', managers', and regulators' best efforts to reduce it further. They are human, too, of course. It is vital that an emphasis on error reduction be maintained, particularly at the early stages of tracking trends, to ensure that the Learning Curve is followed in an orderly way.

Staff at all levels, from individual workers or employees to safety managers, executives, and corporate boards, need to take specific action based on the trends in learning.

Thus, the base contribution of *purely* human error to the rate becomes more significant as we approach the minimum, $A_M$. Typically, human error is estimated to be the major contributor, being between 30 and 90% (and we believe perhaps 100%) of the irreducible error rate. Management systems, safety training, technology improvements, automated systems, and rigorous procedures reduce the rate toward the minimum but do not actually eliminate it.

## Generalized Reporting Format and the Universal Learning Curve

The error, event, or accident rate falls as learning occurs, and the rate of decline is in direct proportion to the instantaneous rate. Thus the general form of the data plot looks like Figure A.1.

There is the initial "startup" transient, as the data are first accumulated and the recorded experience base is very small. The initial rate may seem very high. Once experience is gained (time passes), the expected trend is of a reduction, the slope of which depends on the measure used, the system under study, and the operating goals and standards adopted.

Finally, the rate drops toward an irreducible minimum. The rate may increase or deviate from a steady decline at any time due to:

(a) Insufficient corrective measures
(b) Inadequate standards in goals, procedures, or performance
(c) Changes in the system, operation or technology
(d) Attainment of the minimum rate

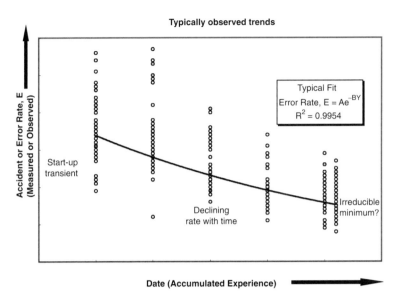

*Figure A.1  Typically observed trends.*

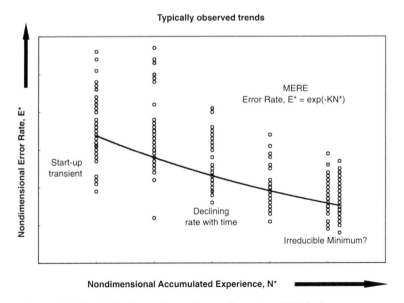

*Figure A.2  Typically observed trends: nondimensional ULC plot.*

Therefore, it is possible to renormalize at any time when necessary or desired (according to equation (9)), and to restart or recommence the plotting and develop a new trend curve. Thus the nondimensional ULC plot form should be used of the form E* (or A*) versus N* (or N), as shown, with the ordinate and abscissa axes scaled from 0 to 1 (Figure A.2).

Practically speaking, we can derive from inspection whether the data are following a well-behaved pattern, have attained a minimum or not, exhibit fluctuating rates, or have adverse increasing error rate trends.

## FORGETTING THEORY

Now we have the MERE written as the rate of decrease of the error rate as proportional to the rate of errors:

$$dA/d\tau = -k \, (A - A_M) \tag{21}$$

where A is the error rate and k is a constant.

We now write $k \equiv L$, the Learning rate (time) constant. In a similar manner, we may include Forgetting by assuming there is a "forgetting" rate constant, F. Hence we write the MERE as the summation of or difference between Forgetting and Learning as follows:

$$dA/d\tau = -L(A - A_M) + F(A - A_M) \tag{22}$$

In terms of observed accident or error rates, A, per unit interval as usually reported in data files, we have, from (6) and (22), the equivalent forgetting solution to the MERE written as

$$(1 - A/A_M) = (1 - A_0/A_M) \exp(-L(1 - F/L)T) \tag{23}$$

So for no forgetting, F = 0, and we revert to our previous Universal Learning Curve. In estimating the accumulated experience, this is in practice related to the total amount of recorded or accumulated experience, $N_T$.

Thus, we may again nondimensionalize the actual accumulated experience, $N(\tau)$, of any experience interval, $\tau$, by writing the accumulated timescale, T, in terms of the fraction of the total accumulated experience, $N_T$, as follows:

$$L(1 - F/L)T = L(1 - F/L)N(\tau)/N_T = L(1 - F/L)N^*, \text{ say,}$$

where $N_T = \Sigma N$, and $N^* \rightarrow 0$ for no experience.

Hence (23) becomes

$$(1 - A/A_M) = (1 - A_0/A_M) \exp - (L(1 - F/L)N^*) \tag{24}$$

so that as $N^* \rightarrow \, ^\circ$ then $A \rightarrow A_M$, as before *the asymptotic minimum error rate.*

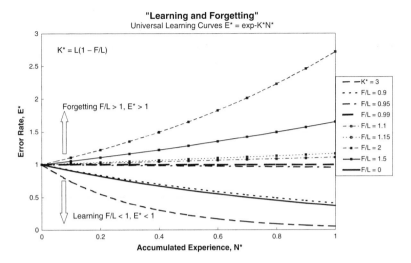

*Figure A.3  Learning and Forgetting.*

This equation (24) is the fundamental useful result of the forgetting analysis, which we will refer to as the *exponential forgetting model* for the error rate as it describes the Universal Forgetting Curves, UFC.

From (24) we have the general MERE form that describes the UFC:

$$(1 - A/A_M) = (1 - A_0/A_M) \exp(-K^*N^*) \tag{25}$$

where $K^* = L(1 - F/L)$, or in the nondimensional form,

$$E^* = \left( \frac{A^*}{A_0^*} \right) = \exp - K^* N^* \tag{26}$$

To a good approximation, $A_0/A_M \gg 1$, and as before we may again write (25) in the simple MERE approximate forgetting form:

$$A = A_M + A_0 \exp(-K^*N^*) \tag{27}$$

Depending on the relative rate of forgetting versus learning, the error rate may decrease or increase, as we have already seen for cases of "insufficient learning."

In Figure A.3, we illustrate the effect of Forgetting on the ULC, and plot E* from (26) for various values for the forgetting to learning ratio, F/L.

Although we have labeled these as Learning Curves, they also may be termed Universal Forgetting Curves. We see immediately that the increasing error rates occur for finite values of F/L > 0, and that the boundary between forgetting, F/L > 1, and learning, F/L, is set by F/L = 1. The dotted line is the best fit to the available world data and implies a value of K* ~ 3. From data we may easily and readily determine the forgetting to learning ratio and use this to numerically assess trends.

# Appendix B

## EXTRACTS FROM EVENT INVESTIGATIONS AND INQUIRIES

### TRAIN ACCIDENT AT LADBROKE GROVE JUNCTION, OCTOBER 5, 1999, THIRD HSE INTERIM REPORT

The accident occurred at 8:09 A.M. on 5 October, when a Thames Train 3-car turbo class 165 diesel unit (the "165" for the purposes of this report) travelling from Paddington to Bedwyn, in Wiltshire collided with a Great Western High Speed Train (the "HST") travelling from Cheltenham Spa to Paddington. The accident took place 2 miles outside Paddington station, at Ladbroke Grove Junction. 31 people died (24 from the 165 and 7 from the HST), with a further 227 taken to hospital. 296 people were treated for minor injuries on site.

- It continues to appear that the initial cause of the accident was that the 165 passed a red signal, SN109, and continued at speed for some 700 m before it collided with the HST. The closing speed was in the region of 145 mph.

- The reasons why the 165 (train) passed the red light are likely to be complex, and any action or omission on the part of the driver *was only one such factor in a failure involving many contributory factors.*

- The driver of the 165 (train) had been assessed as competent in route knowledge, and had obtained his train driver competency certificate 13 days prior to the accident. He had completed 9 shifts as the driver in charge prior to 5 October.

- No evidence has yet been found to indicate that any of the signalling equipment performed otherwise than as expected.

- An initial assessment of the design of the signalling scheme has been carried out and has revealed no significant departures from accepted good practice in terms of the signalling principles that applied at the time the scheme was designed and implemented.

- Extensive examination has been carried out to assess the visibility of signals … we are satisfied that the aspects of those signals relevant to the accident would have been visible for at least the minimum specified periods from the cab of a train travelling at the maximum permitted speed.

- The height of the signals on gantry 8, which carries signal SN 109, is broadly (within 55 mm) in accordance with current requirements regarding the maximum height of signals. At the time the signals were designed and installed no maximum height was specified and subsequent revisions to standards have not required retrospective action.

- The signals on gantry 8 are of unusual design, being in a reverse "L" formation with the red aspect offset to the bottom left. At the time this "L" formation was installed on the gantry (1994) the appropriate standard showed permissible arrangements of aspects in cases where the preferred vertical arrangement could not be accommodated. These did not include the "L" shape or the reverse "L". Furthermore the permissible horizontal arrangements indicated that the red aspect should be closest to the axis of the driver's eye. This is not the case with any of the signals on the gantry. Work is continuing to assess the significance of these factors.

- Due to low bridges, the design and positioning of the overhead electrification equipment and the curvature of the track, the approach view of many of the signals in the Ladbroke Grove area, particularly those on gantry 8, is complex. The human factor investigations have concluded that there are a number of interrelated factors relevant to the potential for driver error on approaching SN109.

- Parts of the Automatic Warning System (AWS) equipment on the HST were severely damaged in the collision. Significantly though the AWS isolation switch, which was removed from the engine compartment, was sealed in the "on" position. Testing of recovered parts of the HST's AWS system identified no faults that could have caused a wrong side failure had they been present prior to the collision.

- The components of the AWS system recovered from the 165 were also extensively damaged, and only limited testing was possible. However, those components, which could be tested, were found to be generally compliant with specifications. In addition, a statement given by the previous driver to drive the train from the same leading end indicated that the AWS was then working normally. Maintenance records indicated that no previous faults with the equipment had been reported.

- Examination of the track following the accident revealed a track joint with significant mis-alignment between the rails in the vicinity of the AWS magnet for SN109. Further work is in progress to assess whether there is any possibility, however remote, that the train passing over this track joint could have caused sufficient vibration of the AWS receiver to cause its mal-operation. If this had happened, it is conceivable that such a mal-operation could have caused the driver of the 165 to receive a false 'green signal ahead' indication as he approached signal SN 109. Although the AWS receiver from the 165 was severely damaged in the accident an identical unit from the same batch, and likely to have similar vibration/shock susceptibility characteristics, is now being tested.

- The "black box" recorder … suffered damage in the collision … its data … analysis identified six significant areas of data (which equate to separate parts of the journey) where anomalies were present and this information can only be considered as low confidence data. The data from the recorder therefore has to be viewed with caution unless backed up by other information such as from the rear recorder data.

- The section of track is equipped with a pilot Automatic Train Protection (ATP) scheme and ATP equipment was fitted to the HST. However, experience has shown the ATP to suffer from reliability problems, and on 5th October the equipment on this HST was isolated because it was not operational. This did not have any bearing on the accident, as any operation of ATP induced emergency braking as the HST went past SN120 would not have materially lessened the speed of impact, which happened shortly after.

- The leading carriage virtually disintegrated in the crash impact. It has now been reconstructed, in two dimensions, from the eighteen rail wagonloads of wreckage removed from the accident site. This is helping to progress the identification of the modes of failure of the vehicle structure. Work is also progressing in correlating fatalities/injuries against the structural damage to the vehicles and comparing the crash performance of aluminum and steel, particularly with respect to the vehicles and fuel tanks of the generation of vehicles involved in the Ladbroke Grove accident.

- An assessment of the on-site fire damage has been completed, and a number of likely sources of ignition identified. Detailed work is continuing to ascertain how the diesel fuel from both trains behaved during the collision, and how it subsequently contributed to the development and spread of the fire.

- SN109 is one of the many signals that is required to be fitted with a train protection system, such as the Train Protection Warning System (TPWS), by 31 December 2003 by virtue of the Railway Safety Regulations 1999. The accident was likely to have been prevented by TPWS.

- No fault was found with the ATP ground based equipment on the approach to signal SN109. Had ATP been fitted to, and had been operational on, the 165 (train) there would not have been a collision.

The phenomena of drivers passing through red signals, apparently unaware of their error are a known, but comparatively rare phenomena. In the absence of an effective automated system to prevent this type of failure, such as Automatic Train Protection (ATP), reliance upon drivers' correctly interpreting and responding to signals alone results in potential for a residual level of risk.

Human factor experts, who are taking the lead on various human factor aspects of this investigation, are therefore investigating the potential for driver human error. Although their investigations are not complete, a number of conclusions, which are considered to be of relevance to the incident, can be drawn. Further human factor work is also being undertaken on passenger escape/evacuation issues, and the operation of the signalling control centre.

The human factors investigations into the immediate causes of the accident have included small group discussions with 29 train drivers familiar with the route taken by the 165 (train) on 5th October. In addition, an analysis of driver tasks was performed by analysing video recordings of a driver travelling over routes of close approximation to that taken by the 165 (train)—this involved three video cameras, one providing a forward view from the driver's cab, a second giving an in-cab view of the driver's actions, and a third giving a view from a head mounted camera worn by the driver. The investigations are addressing the following issues:

- influences relating to human perception and attention, particularly the amount of time available to attend to (the red) signal SN109, processing of this information, and the scope for divided attention or driver distraction;
- influences relating to human information processing and decision making on the basis of information presented by displays and controls available to the driver;
- the potential for issues of habituation (e.g. familiarity or lack of it) and/or driver fatigue having played a role in the accident; and
- the extent to which SN109 could have contributed to the accident in terms of its design, location, orientation and conspicuity.

*The chance of human error can be considered to be enhanced where drivers have a high level of expectation regarding the likely signal aspects which will be displayed,* and where the position of signals fail to take sufficient account of the characteristics and limitations of human sensory and cognitive processes.

The human factors investigations have concluded *that there are a number of interrelated factors relevant to the potential for driver error:*

- There would seem to be good reason for believing that (the) driver did not pass SN109 simply because he was unaware of its presence. The balance of evidence on this would, therefore, seem to suggest that the basis for the incident was one of misinterpretation of the aspect displayed rather than failure to become aware of the presence of gantry 8 and, by implication, signal SN109.
- The comparatively late sighting of SN109 compared with other signals on gantry eight would seem to have potential to lead drivers to make an early decision about the status of this signal and conclude that it is displaying a proceed aspect on the grounds that its lower aspect(s), red or yellow, is not apparent to them.
- The position of the signal gantry, relative to obstructions within the perceptual field and human perceptual processes, has potential for leading drivers to reach erroneous conclusions on the basis of the available sensory data.
- There does not appear to be any evidence indicating the presence of significant distractions present in (the) driver's perceptual field on the approach to SN109 on the day of the accident; neither is there any record of a distraction emanating from the passenger compartment of the train.
- Within the "window of opportunity" for viewing signals, drivers may have to attend to other track and in-cab displays/controls (such potential tasks include looking at the speedometer, attending to AWS warning, looking at the trackside

maximum speed boards etc.). If these actions were undertaken they would sub-tract from the time available to view signals. However, on the basis of a task analysis, if (the) driver attended to all plausible tasks (i.e. worse case) during the SN109 sighting window it would appear that he still had a minimum of approximately 3.5–6 seconds available for unobstructed viewing and undi-vided attention for viewing SN109. None of the tasks for which allowances have been made could be considered unique requirements of (the) driver.

- The AWS audible and visual warnings do not differentiate between caution-ary and stop aspects (contrary to established human factors advice for effec-tive alarm-systems in general).

- Should there have been a malfunction of the AWS, this could have constituted a significant distraction and/or could have contributed to the confirmation of expectations of a proceed aspect being displayed at SN109.

## WALKERTON INQUIRY TRANSCRIPT, DECEMBER 20, 2000, EXAMINATION-IN-CHIEF OF THE GENERAL MANAGER

Q(uestion): And so that we're clear about this, are you telling us that your focus was on the construction main results, rather than results from the drinking water supply for the Town of Walkerton?

A(nswer): My main focus, or only focus was on the 4 and 9 project and how they'd failed.

Q: As the General Manager for the Utility, you bore responsibility for providing safe drinking water for the Town of Walkerton, is that right?

A: That's correct.

Q: So, as far as you were aware then, on Wednesday, May 17th, the new construction main on (the roads) Highway 4 and 9 hadn't been connected yet to the distribution system, is that right?

A: That's correct.

Q: And your focus, you're telling us, was on that unconnected main, rather than on a sample that had to do with the safety of the drinking water supply for five thousand (5000) people, is that what you're saying, sir?

A: That's correct.

Q: I take it though, that because of your concern about the results from the construction project, that when (the testing company) said that he would fax you the results as soon as possible, that caused you to want to have those results to be able to review them as soon as you could, once they came in, is that right?

A: That's correct.

Q: When you spoke to (the testing company), did he say that the water-main samples didn't look good?

A: I think he said that they failed.

Q: What did he say about the distribution 7 system samples?

A: And the distribution samples didn't look good either was the only thing I picked up from the conversation.

Q: And even on hearing that, your focus was still on the samples that related to the unconnected new main, is that right?

A: That's correct.

Q: Why is it that you were so anxious to know about the results from the new main?

A: We were running behind schedule with the Ministry of Transportation's deadline and the contractor was—and they couldn't begin the other contract for the road reconstruction until all the plant or underground work was relocated and changed.

## And a little later in the questioning that same morning:

Q: And this is a Certificate of Analysis in relation to the water that had been sampled from the water supply system on the same day as the water-main samples had been taken, Monday, May 15th 2000, is that right?

A: That's correct.

Q: And this (Certificate) indicates that, in relation to what was purportedly Number 7 treated water and treated water within the distribution system at (two street locations in the town), the samples were positive for both total coli-form and E-coli using the presence/absence method, is that right, sir?

A: That's correct.

Q: In relation to the sample of the Number 7 treated water, there is an indication, and this is the only sample that does have any indication of (contamination) for total coli-form and E-coli as well as HPC, is that right sir?

A: That's correct.

Q: You knew then that CFU stood for colony forming unit, is that right?

A: Not really, sir.

Q: Not really. You knew that at least it was a means of measuring bacteria?

A: Correct.

Q: And did you know that HPC stood for heterotrophic plate count?

A: Yes, I did.

Q: And this Certificate of Analysis indicates that there were greater than two hundred (200) CFU's of total coli-forms for one hundred (100) millilitres of water, is that right?

A: That's correct.

Q: It indicates that there were greater than two hundred (200) CFU's of E. coli per 100 millilitres of water, is that right?

A: That's correct.

And a little later in the same dialog:

Q: Do you agree with me, that the results shown in that Certificate of Analysis, point to a very serious problem with the water being supplied to the people of Walkerton?

A: Yes.

Q: And you knew that at the moment you first looked at this; is that right, sir?

A: Yes, I did.

Q: Now I understand that you worked on Thursday, May 18th 2000?

A: Yes.

Q: Did you have a conversation that day with (one of your staff), about installation of a chlorinator at Well 7?

A: Yes, I did.

Q: By the way, anyone with knowledge of that Certificate of Analysis that we've been discussing that pointed to a serious problem with the Walkerton Water Distribution System and water supply, would be very concerned, I suggest, about there being no chlorinator at Well 7; is that right?

A: If I had noticed it Thursday.

Q: What did (your staff) tell you in that respect?

A: That they were working on it and he assured me that it would be up and running by Friday."

# Appendix C

## REFERENCES, BIBLIOGRAPHY, AND SOURCES

The following is a partial listing of the primary sources and references we have used. The list is arranged by chapter or chapter section where the references are first introduced in the text. The list also includes sources that the reader may find useful and/or interesting to consult. These are correct to the best of the authors' knowledge and belief, but the sources and cited data files may not be active, or they may contain changed, revised, or updated data files or website content or information. Therefore, although we have made every effort to be exact, the authors cannot be held responsible for any incorrect or inaccurate information.

### CHAPTER 1: MEASURING SAFETY

Airsafe.com (2000). http://www.airsafe.com, "Fatal Events and Fatal Event Rates from 1970–2000."

Allison, P.D. (1984). *Event History Analysis.* Sage Publications, Thousand Oaks, CA.

Amalberti, R. (2001). "The Paradoxes of Almost Totally Safe Transportation Systems." *Safety Sciences* **37**, 109–126.

BTS (Bureau of Transportation Safety) (2000). "Incident Data and Transportation Statistics." U.S. Department of Transportation.

Denham, T. (1996). *World Directory of Airliner Crashes.* Patrick Stephens Ltd., Haynes Publishing, U.K.

Duffey, R.B., and Saull, J.W. (2000). "Error and Accident Rates." In: Proceedings of the 8th International Conference on Nuclear Engineering, ICONE- 8, Baltimore, MD.

Duffey, R.B., and Saull, J.W. (2001). "Errors in Technological Systems." In: Proceedings, World Congress, Safety of Modern Technical Systems, Saarbrucken, Germany.

Graunt, J. (1662). "Foundation of Vital Statistics." In: *The World of Mathematics* (J.R. Newman, ed.), Vol. 3, p. 1421. Simon and Schuster, New York, 1956.

Halley, E. (1693). "An Estimate of the Degrees of Mortality of Mankind." *Philosophical Transactions of the Royal Society* XVII. Also in: *The World of Mathematics* (J.R. Newman, ed.), Vol. 3, p. 1437. Simon and Schuster, New York, 1956.

Howard, R.W. (1991). "Breaking Through the $10^6$ Barrier." In: Proceedings of the International Federation of Airworthiness Conference, Auckland, New Zealand.

Ott, K.A., and Campbell, G. (1979). "Statistical Evaluation of Major Human Errors During the Development of New Technological Systems." *Nuclear Science and Engineering* **71**, 267–279.

Ott, K.A., and Marchaterre, J.F. (1981). "Statistical Evaluation of Design-Error Related Nuclear-Reactor Accidents." *Nuclear Technology* **52**, 179.

U.S. Department of Transportation National Highway Traffic Safety Administration (DOT NHTSA) (1997). "Traffic Safety Facts 1997."

## Chapter 2: Traveling in Safety

### Air

AirClaims Ltd. (2000). "JAA and North America Fatal Accident Data." Author communication.

Airsafe Financial (1998). "Fatal Events and Fatal Event Rates by Airline Since 1970" (http://www.airsafe.com/airline.htm) and Airsafe.com (2000). "Fatal Events by Airline" (http://www.airsafe.com), 01.09.00.

BEA (2002). "Accident on 25 July 2000 at La Patte d'Oie in Gonesse (95) to the Concorde registered F-BTSC operated by Air France." Rapport du Bureau Enquets-Accident (BEA), Ministère de l'équipement, des Transports et du Logement, Inspection Générale de l'Aviation Civile, France, January Report translation f-sc000725a.

BTS (Bureau of Transportation Safety) (2000). "Number of Pilot-Reported Near Midair Collisions (NMAC) by Degree of Hazard," 1980–1997, Table 3-16. U.S. Department of Transportation.

CAA (U.K. Civil Aviation Authority) (2000). "Aviation Event Data." Author communication.

*Conde Nast Traveler* (1995). "Accident Rates of the World's 85 Major Airlines." December.

Duffey, R.B., and Saull, J.W. (1999). "On a Minimum Error Rate in Complex Technological Systems." In: Proceedings of the 52nd Annual International Air Safety Seminar on Enhancing Safety in the 21st Century, Rio de Janeiro, Brazil, pp. 289–305.

FAA (U.S. Federal Aviation Agency) (1999). "System Indicators." U.S. Department of Transportation (U.S. DOT) (see also www.asy.faa.gov/safety_analysis/Si.htm).

Reason, J. (1990). *Human Error.* Cambridge: Cambridge University Press.

Russell, P. (1998). "Aviation Safety—The Year in Review." Boeing Commercial Airplane Group paper at the Joint International Meeting, FSF/IFA/IATA. "Aviation: Making a Safe System Safer," 51st Flight Safety Foundation International Air Safety Seminar. Cape Town, South Africa, November 17–19.

Saull, J.W., and Duffey, R.B. (2000). "Aviation Events Analysis." In: Proceedings, Joint Meeting of the Flight Safety Foundation (FSF) 53rd Annual International Air Safety Seminar (IASS), International Federation of Airworthiness (IFA) 30th International Conference and International Air Transport Association (IATA), "Improving Safety in a Changing Environment," New Orleans, LA, October 29–November 1.

TSB (Transport Safety Board) (2000, 1997). *Statistical Summary of Aviation Occurrences.* Canada.

U.K. Airprox Board (2000). Author communication.

### Sea

Berman, B.D. (1972). *Encyclopaedia of American Shipwrecks.* Mariners Press, Boston.

Hocking, C. (1969). *Dictionary of Disasters at Sea during the Age of Steam, 1824–1962.* Compiled and published by Lloyd's Register of Shipping (see also http://www.lr.org).

Institute of London Underwriters (1988 et seq.). "Hull Casualty Statistics," data for 1987–1997, International Union of Marine Insurance Conferences (see also http://www.iua.co.uk).

MAIB (U.K. Marine Accident Investigation Branch) (2000). U.K. DETR (Department of Environment, Transport and the Regions).

McCluskie, T., Sharpe, M., and Marriott, L. (1999). *Titanic and Her Sisters.* Prospero Books, London.

Parsons, R.C. (2001). *In Peril on the Sea.* Pottersfield Press, Nova Scotia, Canada.

Pomeroy, V. (2001). "Classification—Adapting and Evolving to Meet Challenges in the New Safety Culture." *Safety of Modern Technical Systems.* TüV, Saarland Foundation, Germany, p. 281.

Portsmouth Naval Museum (1999). Mary Rose Ship Exhibit (see also http://www.maryrose.org).

Ritchie, D. (1996). *Shipwrecks.* Checkmark Books, New York.

U.K. Protection and Indemnity Mutual Insurance Club (2000). "Analysis of Major Claims," London (see http://www.ukpandi.com).

Vasa Museet (1998). Vasa Ship Exhibit, Stockholm (see also http://www.vasamuseet.se/).

### Road

Australian Transport Safety Bureau (2001). "Road Fatalities Australia 2000 Statistical Summary."

Continental Tires, Corporate News (2001). July 24. (http://www.contigentire.com/news.cfm).

Firestone Tire Company (2000). "Firestone Tire Recall" (see also http://www.Firestone.com).

Goodyear Tire Company (2000). "The Future of Tire Safety." Joseph M. Gingo, Global Automotive Safety Conference, Automotive Div. Society of Plastics Engineers Michigan State University Executive Education Center, Troy, MI (see also http://www.goodyear.com).

Lampe, John L. (2001). "Testimony of John Lampe." U.S. House Committee on Energy and Commerce, Subcommittee on Commerce, Trade and Consumer Protection, June 19. "Hearing on the Ford Motor Company's Recall of Certain Firestone Tires," p. 212 (from http://www.house.gov/commerce/).

NHTSA (National Highway Transport Safety Administration) (1997). "Traffic Safety Facts." U.S. Department of Transportation.

NHTSA (National Highway Transport Safety Administration) (2001). http://www.nhtsa.dot.gov/hot/firestone/Update.html.

U.K. (1998). *Transport Statistics Great Britain 1997.* Tables 8.3 and 8.7, International Comparisons. HMSO, U.K.

U.K. DETR (Department of Environment, Transport and the Regions) (2000). "Road Accident Casualties by Road User Type and Severity: 1988–1998" and "International Comparisons of Transport Statistics, 1970–1994."

UN/ECE, Transport Division Road Safety, data files.

U.S. Department of Transportation (U.S. DOT) (1999). Bureau of Transportation Statistics (BTS) on National Transportation Statistics, Table 3-33 (see also http://www.bts.gov/ntda/nts/).

U.S. Department of Transportation (U.S. DOT), National Highway Traffic Safety Administration (NHTSA) (2001). "Engineering Analysis Report and Initial Decision Regarding EA00-023: Firestone Wilderness AT Tires." Safety Assurance, Office of Defects Investigation, October.

*Rail*

AAR (Association of American Railroads) (1999). "U.S. Railroad Safety Statistics and Trends," by Peter W. French. September 20.

EU (1999). European Transport Safety Council, "Exposure Data for Travel Risk Assessment: Current Practice and Future Needs in the EU." June, Brussels.

FRA (U.S. Federal Railroad Administration) (2000). "Operational and Accident Data for 1975–2000" (see also http://safetydata.fra.dot.gov/officeofsafety/).

H&SE (2000). Railways Inspectorate, "Train Accident at Ladbroke Grove Junction, 5 October 1999, Third HSE Interim Report." U.K. Health and Safety Executive.

H&SE (2000). Railways Inspectorate, "Train Derailment at Hatfield, 17 October 2000, First HSE Report." U.K. Health and Safety Executive.

HM Railway Inspectorate (2000). Health and Safety Executive, "History of SPADs on Railtrack's Controlled Infrastructure (RCI)" (see also http://www.hse.gov.uk/railway/rsb9798.htmTable).

HM Railway Inspectorate (2000). Health and Safety Executive, "Railway Safety Statistics" (see also http://www.hse.gov.uk/railway/).

Statistics Canada (1999). "Annual Railroad Ten-Year Trends." Cat. 52-216.

U.K. Health and Safety Commission (2001). *The Ladbroke Grove Rail Inquiry*, Part 1 Report, Lord Cullen. HSE Books, HMSO, Norwich, U.K.

U.K. Health and Safety Executive (2001). *The Southall and Ladbroke Grove Joint Inquiry into Train Protection Systems*, John Uff and Lord Cullen. HSE Books (see also http://www.hse.gov.uk/railway/spad/lgri1.pdf).

## Chapter 3: Working in Safety

Amalberti R. (2001). "The Paradoxes of Almost Totally Safe Transportation Systems." *Safety Science* 37, 109–126.

Baumont, G., Bardou, S., and Matahri, N. (2000). "Indicators of Plants Performance During Events Identified by Recuperare Method." Specialist Meeting on Safety Performance Indicators, Madrid, Spain, October 17–19, 2000.

BLS (U.S. Bureau of Labor Statistics) (1999). "Tables of Incidence Rates of Occupational Injuries by Industry." U.S. Department of Labor, OSHA, Bureau of Labor Statistics (see also http://www.bls.gov/specialrequests/ocwc/oshwc/osh/os/ost60641).

BLS (U.S. Bureau of Labor Statistics) (2000). "Fatal Workplace Injuries 1974–1991." Department of Labor, OSHA, Bureau of Labor Statistics (http://www.osha.gov/osstats/bls/priv0.html and priv9 and privtbl).

Bull, D.C., Hoffman, H.J., and Ott, K.O. (1984). "Trend Analysis of Boiler Explosions Aboard Ships." Personal communication of report.

Campbell, G., and Ott, K.O. (1979). "Statistical Evaluation of Major Human Errors During the Development of New Technological Systems." *Nuclear Science and Engineering* 71, 267.

Chamber of Mines of South Africa (2001). "Mining Data Library: Mine Safety Statistics" (see also http://www.bullion.org.za/bulza/stat/Safty/safstat.htm).

Davies, R., and Elias, P. (2000). "An Analysis of Temporal and National Variations in Reported Workplace Injury Rates." UK Health and Safety Executive, University of Warwick, Institute for Employment Research.

Duckham, H., and Duckham, B. (1973). *Great Pit Disasters*. Chapter 9. David & Charles, Newton Abbott.

Joint Federal–Provincial Inquiry Commission into Safety in Mines and Mining Plants in Ontario (Canada) (1981). "Towards Safe Production." Report of the Joint Federal–Provincial Inquiry Commission into Safety in Mines and Mining Plants in Ontario ("The Burkett Report").

*Journal of Mines, Metals and Fuels* (2001). "Special Issue on Mine Safety." Vol. XLIX.

Kohda, T., Nojiri, Y., and Inoue, K. (2000). "Root Cause Analysis of JCO Accident Based on Decision-Making Model." Proceedings PSAM5, Probabilistic Safety Assessment and Management, November 27, Osaka, Japan.

LANL (Los Alamos National Laboratory Report) (2000). "A Review of Criticality Accidents." LA 13638, 2000 revision.

MITI (2000), author communication.

Perrow, C. (1984). *Normal Accidents: Living with High-Risk Technologies.* Basic Books, New York.

Reason, J. (1990). *Human Error.* Cambridge University Press, Cambridge.

Queensland Department of Mines and Energy Safety, Division of Mines Inspectorate (2001). *Fatalities in the Queensland Coal Industry 1882–1999* (see also http://www.dme. qld.gov.au/safety/index.htm).

Takala, J. (1998). "Global Estimates of Fatal Occupational Accidents." International Labor Office, United Nations ILO, 16th International Conference of Labour Statisticians, Geneva, 6–15th October, Paper ICLS/16/RD 8 (see also: http://www.ilo.org).

U.K. Chemical Industries Association (CIA) (1998). "Safety Award Scheme—Annual Survey and Statistics," RC103 (see also http://www.cia.org.uk/).

U.K. Department of Trade and Industry (DTI) (2001). "The Government's Expenditure Plans 2001–02 to 2003–04 and Main Estimates 2001–02," Cm 5112, Section 8.33-34, "Issuing Import Licenses," p. 115 (see also http://www.dti.gov.uk/expenditureplan/ expenditure2001/pdfs/pdfs/full.pdf).

U.K. Safety Assessment Federation (2001).

U.S. Department of Labor (1998). Occupational Safety & Health Administration (OSHA). "Workplace Injury, Illness and Fatality Statistics" (see also http://www.osha.gov/ oshstats/work.html).

U.S. Department of Labor (2001). Mine Safety and Health Administration (MSHA). "Statistics: Historical Data" (see also http://www.msha.gov).

World Association of Nuclear Operators (WANO) (1999). "WANO Performance Indicators." 1998 (http://www.nei.org/library/facts.html).

## CHAPTER 4: LIVING IN SAFETY

AMA (American Medical Association) (2000). National Patient Safety Foundation (http://www.ama-assn.org/med-sci/npsf/research).

Aviation Safety Network (2001). Hijack Database (http://aviation-safety.net/index.shtml).

Boeing (2001). "Statistical Summary of Commercial Jet Airplane Accidents World-wide Operations 1959–1999." Airplane Safety, Boeing Commercial Airplanes Group, Seattle, WA.

Breyer, S. (1993). *Breaking the Vicious Circle.* Harvard University Press, Cambridge. MA.

Canadian Institute for Health Information (CIHI) (2000). "Health Indicators." Statistics Canada, Catalogue Number 82-221-XIE.

CDC National Center for Health Statistics (2001). "National Vital Statistics Reports" (see also http://www.cdc.gov/nchs/fastats/deaths.htm).

CIHI (2001). Canadian Institute for Health Information. "Health Indicators December 2000 Report." Catalogue #82-221-XIE (see also http://www.cihi.ca).

Craun, G.F. (1991). "Causes of Waterborne Outbreaks in the United States." *Water Science* **24**(2), 17–20. (Also U.S. Centers for Disease Control, 1991–1998, "Surveillance of Waterborne Disease Outbreaks," CDC Morbidity and Mortality Weekly Reports (1993), pp. 1–22, 42(ss-05); (1996) 45(ss-01); (1998) pp. 1–34, 47(ss-05); (2000) pp. 1–35, 9(ss-04).

EU (2000). "Report of the International Task Force for Assessing the Baia Mare Accident." The Baia Mare Task Force, Tom Garvey, Chairman, December, Brussels, Belgium.

Federal Emergency Management Agency (2001). "Dam Safety: Progress through Partnerships," and "U.S. Dam Safety Program," U.S. FEMA (see also http://www.fema.gov/mit/damsafe/news/news0008.htm).

HRDC (Human Resource and Development Commission) (2001). "Occupational Injuries and Their Cost by Province or Territory 1993–1997." Canada.

IOM (2000). *To Err Is Human: Building a Safer Health System* (L.T. Kohn, J.M. Corrigan, and M.S. Donaldson, eds.). Committee on Quality of Health Care in America, U.S. National Academy's Institute of Medicine, National Academy Press, Washington, D.C. (also http://www.Aetnaushc.com. "Aetna U.S. Health Care Announces New Initiatives to Help Reduce Medical Errors" and *British Medical Journal*, March 18, 320(7237)).

*Maclean's Magazine* (2001). "Mistakes That Kill." Vol. 114, No. 3, pp. 38–43, August.

McKeown, T. (1978). "Determinants of Health in Human Nature." April (also quoted in Horwitz, L. and Ferleger, L., *Statistics for Social Change.* Black Rose Books, Montreal, Canada, 1980).

National Performance of Dams Program (2001). (http://npdp.stanford.edu/).

O'Connor, D.R. (2002). Walkerton Inquiry, Final Report, Part One, "The Events of May 2000 and Related Issues." January 18.

Ott, K.O., Hoffman, H.-J., and Oedekoven, L. (1984). *Statistical Trend Analysis of Dam Failures since 1850.* Kernforschungsanlage Julich GmbH (KFA), Jul-Spez-245.

Pettijohn, S. (1985). "Case Study Report on the Therapy Mis-administrations Reported to the NRC Pursuant to 10 CFR 35.42." U.S. Nuclear Regulatory Commission, AEOD/C505.

Phillips, D.P., Christenfeld, N., and Glynn, L.M. (1998). "Increase in U.S. Medication-Error Deaths between 1983 and 1993." *The Lancet* 351, 643.

Statistics Canada (2002). Population data files (http://www.statcan.ca/english/Pgdb/People/Population/demo31a.htm).

Tatolovich, J. (1998). "Comparison of Failure Modes from Risk Assessment and Historical Data for Bureau of Reclamation Dams." January. Materials Engineering and Research Laboratory DSO-98-01, U.S. Department of the Interior, Bureau of Reclamation, Dam Safety Office.

U.K. Secretary of State for Health (2001). CM-5207, Report of the Public Inquiry into children's heart surgery at the Bristol Royal Infirmary 1984–1995. "Learning from Bristol," 2001, Chair Professor Ian Kennedy, Full Report and Annex A INQ. Papers on statistical analysis prepared for the Inquiry (see also http://www.bristol-inquiry.org.uk).

U.S. Army Corps of Engineers (2001). "National Inventory of Dams" (see also http://crunch.tec.army.mil/nid/webpages/nid.cfm).

U.S. Nuclear Regulatory Commission (1997). "Assessment of the Quality Management Program and Misadministration Rule," SECY-97-037, February 12 (see also USNRC AEOD NUREG Report-1272, "Nuclear Materials Annual Report 1994–1995," September 1996).

Walkerton Inquiry (2001). "The Economic Costs of the Walkerton Water Crisis." Commission Paper 14, J. Livernois, Toronto.

Walkerton Inquiry Transcripts (2000–2001). Examination-in-Chief of the General Manager, December 10th, 2000, and Testimony of the Premier of Ontario, June 29th, 2001. "The Walkerton Inquiry," The Honourable Dennis R. O'Connor, Commissioner (http://www.walkertoninquiry.com/index.html).

## CHAPTER 5: ERROR MANAGEMENT: STRATEGIES FOR REDUCING RISK

Barrow, John D., and Tipler, Frank J. (1988). *The Anthropic Cosmological Principle.* Oxford University Press, New York.

Breyer, S. (1993). *Breaking the Vicious Circle.* Harvard University Press, Cambridge, MA.

Brock, J.A. (1998). DuPont Safety and Environmental Management Services.

Crosby, Philip B. (1984). *Quality Is Free.* Chapter 8, "The Third Absolute: The Performance Standard Is Zero Defects." McGraw-Hill Book Company, New York.

Dorner, D. (1996). *The Logic of Failure.* Chapter 2. Perseus Books, Cambridge, MA.

Duane, J.T. (1964). "Learning Curve Approach to Reliability Monitoring." *Trans. IEEE on Aerospace* 2(2), 563–566.

Howard, R.W. (2000). "Planning for Super Safety: The Fail-safe Dimension." *The Aeronautical Journal of the Royal Aeronautical Society*, November, 517–555.

Perrow, C. (1984). *Normal Accidents: Living with High-Risk Technologies.* Basic Books, New York.

Rasmussen, J. (1986). *Information Processing and Human-Machine Interaction.* North-Holland, Amsterdam.

Trimtop, R.M. (1990). "Risk-taking Behaviour: Development and Empirical Examination of Risk Motivation Theory." Ph.D. Thesis, Queen's University, Kingston, Ontario, Canada.

Von Neumann, John (date unknown). "General and Logical Theory of Automata." In: *The World of Mathematics* (J.R. Newman, ed.). Simon and Schuster, New York, 1956.

Wilde, G.S. (1989). "Accident Countermeasures and Behavioral Compensation: The Position of Risk Homeostasis Theory." *Journal of Occupational Accidents* 10, 267–292.

## APPENDIX A: UNIVERSAL LEARNING CURVES AND THE EXPONENTIAL MODEL

Stock, D., Veseley, W., and Samanta, P. (1994). "Development and Application of Degradation Modeling to Define Maintenance Practices." U.S. Nuclear Regulatory Commission Report NUREG/CR-5967, June.

Tobias, P.A., and Trindade D.C. (1986). *Applied Reliability.* Van Nostrand Reinhold, New York.

# NOMENCLATURE

| | |
|---|---|
| A | Rate of errors, events, or accidents |
| A* | Nondimensional rate, $(1 - A/A_M)$ |
| $A_0$ | Initial rate |
| $A_0^*$ | Nondimensional rate, $(1 - A_0/A_M)$ |
| accN | Accumulated experience |
| $A_M$ | Minimum rate |
| AR | Accumulated rate, A/accN |
| CR | Constant rate |
| CR(0) | Initial recorded rate at $N_0$ |
| E | Number of events or errors |
| E* | Nondimensional error rate, $A^*/A_0^*$ |
| F | Forgetting rate constant |
| F(t) | Failure probability |
| IR | Instantaneous rate, A/N |
| k | Learning rate constant |
| K* | Nondimensional learning rate, $L(1 - F/L)$ |
| L | Learning rate constant |
| M | Millions |
| m | Number of reporting intervals |
| N | Experience measure |
| N* | Nondimensional accumulated experience, $(N/N_T)$ |
| $N_0$ | Initial experience at CR(0) |
| $N_T$ | Total experience |
| t | Time |
| $\tau$ | Instant of accumulated experience |
| $\lambda$ | Failure rate |

# INDEX

Page references followed by "t" denote tables; "f" denote figures; common organizational or usual working abbreviations are in parentheses.